Frauke Marie Schnorfeil

MicroRNA Regulation of Dendritic Cells

Frauke Marie Schnorfeil

MicroRNA Regulation of Dendritic Cells

Experimental studies on the role of microRNAs in dendritic cell development and function

Südwestdeutscher Verlag für Hochschulschriften

Impressum/Imprint (nur für Deutschland/only for Germany)
Bibliografische Information der Deutschen Nationalbibliothek: Die Deutsche Nationalbibliothek verzeichnet diese Publikation in der Deutschen Nationalbibliografie; detaillierte bibliografische Daten sind im Internet über http://dnb.d-nb.de abrufbar.
Alle in diesem Buch genannten Marken und Produktnamen unterliegen warenzeichen-, marken- oder patentrechtlichem Schutz bzw. sind Warenzeichen oder eingetragene Warenzeichen der jeweiligen Inhaber. Die Wiedergabe von Marken, Produktnamen, Gebrauchsnamen, Handelsnamen, Warenbezeichnungen u.s.w. in diesem Werk berechtigt auch ohne besondere Kennzeichnung nicht zu der Annahme, dass solche Namen im Sinne der Warenzeichen- und Markenschutzgesetzgebung als frei zu betrachten wären und daher von jedermann benutzt werden dürften.

Coverbild: www.ingimage.com

Verlag: Südwestdeutscher Verlag für Hochschulschriften GmbH & Co. KG
Heinrich-Böcking-Str. 6-8, 66121 Saarbrücken, Deutschland
Telefon +49 681 37 20 271-1, Telefax +49 681 37 20 271-0
Email: info@svh-verlag.de

Approved by: Munich, LMU, Dissertation, 2011

Herstellung in Deutschland:
Schaltungsdienst Lange o.H.G., Berlin
Books on Demand GmbH, Norderstedt
Reha GmbH, Saarbrücken
Amazon Distribution GmbH, Leipzig
ISBN: 978-3-8381-3189-4

Imprint (only for USA, GB)
Bibliographic information published by the Deutsche Nationalbibliothek: The Deutsche Nationalbibliothek lists this publication in the Deutsche Nationalbibliografie; detailed bibliographic data are available in the Internet at http://dnb.d-nb.de.
Any brand names and product names mentioned in this book are subject to trademark, brand or patent protection and are trademarks or registered trademarks of their respective holders. The use of brand names, product names, common names, trade names, product descriptions etc. even without a particular marking in this works is in no way to be construed to mean that such names may be regarded as unrestricted in respect of trademark and brand protection legislation and could thus be used by anyone.

Cover image: www.ingimage.com

Publisher: Südwestdeutscher Verlag für Hochschulschriften GmbH & Co. KG
Heinrich-Böcking-Str. 6-8, 66121 Saarbrücken, Germany
Phone +49 681 37 20 271-1, Fax +49 681 37 20 271-0
Email: info@svh-verlag.de

Printed in the U.S.A.
Printed in the U.K. by (see last page)
ISBN: 978-3-8381-3189-4

Copyright © 2012 by the author and Südwestdeutscher Verlag für Hochschulschriften GmbH & Co. KG and licensors
All rights reserved. Saarbrücken 2012

Table of contents

1 Summary 3

2 Zusammenfassung 5

3 Aims of this work 7

4 Introduction 9

   4.1 Dendritic cells 9

      4.1.1 DC subsets 10

         4.1.1.1 $CD11b^{+}$-like and $CD8\alpha^{+}$-like cDCs 10

         4.1.1.2 Langerhans cells 11

         4.1.1.3 pDCs 12

         4.1.1.4 Inflammatory DCs 13

      4.1.2 DC development 13

         4.1.2.1 Cytokines 14

         4.1.2.2 Transcription factors 14

      4.1.3 DC activation and maturation 15

      4.1.4 Antigen presentation and T cell activation 15

      4.1.5 DCs in clinical therapy 16

         4.1.5.1 DC vaccination 17

         4.1.5.2 In vivo DC targeting 17

         4.1.5.3 Novel strategies in DC-based tumortherapy 18

   4.2 miRNAs 20

      4.2.1 miRNA biogenesis and target recognition 20

      4.2.2 Regulation of miRNA expression 21

      4.2.3 miRNA regulation of gene expression 21

   4.3 miRNAs in the immune system 23

      4.3.1 miRNA regulation of hematopoietic stem cell differentiation 23

      4.3.2 miRNA regulation of lymphocyte development and function 23

    4.3.3   miRNA regulation of innate inflammatory responses 24

    4.3.4   miRNAs in hematopoietic malignancies 25

5   Publication I (Kuipers et al., Mol Immunol 2010) 27

6   Publication II (Kuipers et al., J Immunol 2010) 37

7   Discussion 49

    7.1   miRNAs regulate DC differentiation 50

        7.1.1   miRNAs regulate DC lineage and subset differentiation by targeting Notch/Wnt signaling 50

        7.1.2   miRNAs influence DC subset differentiation by controlling the expression of transcription factors 51

    7.2   miRNAs regulate DC survival and function 57

        7.2.1   miRNAs regulate DC turnover and apoptosis 57

        7.2.2   miRNAs fine-tune DC maturation and function 59

            7.2.2.1   miRNAs regulate DC antigen presentation 60

            7.2.2.2   miR-146a and miR-155 are key miRNAs in DC maturation 62

    7.3   Outlook – Implications for immunotherapy 65

8   References 67

9   List of abbreviations 79

1 Summary

Dendritic cells (DCs) play a key role in the initiation of adaptive immune responses and the maintenance of self-tolerance. Due to their therapeutic potential, understanding the mechanisms that guide DC differentiation and effector functions is important. DC differentiation and activation depends on transcription factor control of stage-specific gene expression. The recent identification of posttranscriptional control of gene expression by microRNAs (miRNAs) has added another layer of gene regulation that might be important in DC biology.

We analyzed the miRNA expression profiles of different DC subsets and identified several miRNAs differentially expressed between plasmacytoid DCs (pDCs) and conventional DCs (cDCs). In terms of miRNA expression, pDCs were more closely related to $CD4^+$ T cells than to cDCs. We also observed that pDCs and cDCs preferentially expressed miRNAs associated with lymphoid or myeloid lineage differentiation, respectively. By knocking down miR-221 or miR-222 during *in vitro* DC differentiation, we obtained a higher pDC frequency. While $p27^{kip1}$ and c-kit are confirmed miR-221/222 targets, we additionally identified the pDC cell fate regulator E2-2 as a potential miR-221/222 target. Thus, our analysis points to a role for miRNAs in directing and stabilizing pDC and cDC cell fate decisions.

To assess the general influence of miRNAs on DCs, we generated mice with a DC-specific conditional knockout of the key miRNA-producing enzyme Dicer. Dicer-deficient mice displayed no alterations in short-lived spleen and lymph node DCs. However, long-lived epidermal DCs, known as Langerhans cells (LCs), showed increased turnover and apoptosis rates, leading to their progressive loss. Upon stimulation, Dicer-deficient LCs were able to properly upregulate the surface molecules MHC class I and CCR7, but not MHC class II, CD40 and CD86. In consequence, they were incapable of stimulating $CD4^+$ T cell proliferation.

The work presented here indicates a role for miRNAs in DC regulation not covered by transcription factors. Having demonstrated a role for miRNAs in DC lineage fate decisions, as well as in LC homeostasis, maturation and function, we conclude that miRNAs regulate various aspects of DC biology and thereby contribute to the control of adaptive immune responses.

2 Zusammenfassung

Dendritische Zellen (*dendritic cells*, DCs) haben eine zentrale Funktion sowohl bei der Initiierung von adaptiven Immunantworten als auch bei der Aufrechterhaltung der Selbst-Toleranz. Aufgrund ihres therapeutischen Potentials ist es wichtig die Mechanismen zu verstehen, die die Differenzierung und Effektorfunktionen von DCs steuern. Die DC-Differenzierung und -Aktivierung ist von Transkriptionsfaktoren abhängig, die die phasenspezifische Genexpression kontrollieren. Die kürzlich entdeckte posttranskriptionelle Kontrolle der Genexpression durch microRNAs (miRNAs) stellt eine weitere Ebene der Genregulation dar, die in der DC-Biologie von Bedeutung sein könnte.

Im Rahmen dieser Arbeit wurden miRNA-Expressionsmuster von verschiedenen DC-Subpopulationen analysiert und differentiell exprimierte miRNAs in plasmazytoiden DCs (pDCs) und konventionellen DCs (cDCs) identifiziert. pDCs und cDCs exprimierten vorrangig miRNAs, die mit lymphoider bzw. myeloider Zelldifferenzierung assoziiert werden. Die Inhibierung von miR-221 und miR-222 während der *in vitro* DC-Differenzierung resultierte in einer erhöhten pDC-Frequenz. Dafür verantwortlich sein könnte neben den nachgewiesenen miR-221/222-Zielen $p27^{kip1}$ und c-kit auch der von uns als potentielles Ziel identifizierte Regulator E2-2, der für die pDC-Differenzierung essentiell ist. Somit weist unsere Analyse auf eine Funktion von miRNAs bei der Determinierung und Stabilisierung von pDC- und cDC-Zellidentitäten hin.

Um den generellen Einfluss von miRNAs auf DCs zu untersuchen, wurden Mäuse mit einem DC-spezifischen konditionellen Knockout des für die miRNA-Generierung wichtigen Enzyms Dicer erzeugt. Dicer-defiziente Mäuse wiesen keine Veränderungen der kurzlebigen Milz- und Lymphknoten-DCs auf. Die langlebigen epidermalen DCs, bekannt als Langerhans Zellen (LCs), zeigten jedoch eine erhöhte Erneuerungs- und Apoptose-Rate, was zu ihrem progressivem Verlust führte. Zudem konnten stimulierte Dicer-defiziente LCs zwar die Oberflächenmoleküle MHC Klasse I und CCR7 korrekt hochregulieren, nicht jedoch MHC Klasse II, CD40 und CD86. Infolgedessen zeigten sich LCs ineffizient in der Stimulation von $CD4^+$ T Zellen.

Die hier vorgestellten Ergebnisse weisen auf eine Rolle von miRNAs in der DC-Regulation hin, die nicht von Transkriptionsfaktoren gedeckt wird. Die gezeigte Bedeutung von miRNAs in der Determinierung der DC-Zellidentität, sowie in der Homöostase, Reifung und Funktion von LCs, lässt den Schluss zu, dass miRNAs diverse Aspekte der DC-Biologie regulieren und dadurch zur Kontrolle von adaptiven Immunantworten beitragen.

3 Aims of this work

As Dendritic cells (DCs) are key modulators of adaptive immune responses, tight regulation of their development and function is important. The principle aim of this work was to elucidate whether and how miRNAs contribute to DC regulation.

The differentiation of functionally different DC subsets is known to depend on transcription factors. We sought to assess whether miRNAs also contribute to DC differentiation by analyzing miRNA expression patterns in DC subsets. Following the identification of differentially expressed miRNAs, we aimed to manipulate the expression of these miRNAs and analyze the cellular outcomes with regard to potential miRNA target genes.

In parallel, we wanted to analyze the general impact of miRNAs on DC development and function. To that end, we designed a transgenic mouse model with a DC-specific deletion of the key miRNA-generating enzyme Dicer. Using this model, in which a lack of functional miRNAs in DCs was expected, we aimed to analyze the influence of miRNAs on DC homeostasis, maturation and function.

4 Introduction

4.1 Dendritic cells

Both the innate and adaptive immune systems are vital for maintaining health. Dendritic cells (DCs) are bone marrow (BM)-derived cells that hold a unique function at the interface of both arms of immunity. Widely distributed in the body, they are specialized in the recognition and uptake of pathogens, the activation of natural killer (NK) cells, the processing and presentation of antigens and consequently in the induction of adaptive immune responses via B and T cells (reviewed in Steinman and Idoyaga, 2010).

A general model of DC function is the 'Langerhans cell paradigm', named after a DC subset found in the skin (reviewed in Wilson and Villadangos, 2004). According to this model, DCs reside in an immature state in peripheral tissues, where they scan their environment for 'danger signals'. In this resting state they express various receptors for pathogen recognition. Upon pathogen encounter, signaling through these receptors activates the DCs, initiating their migration to the draining lymph nodes (LNs). During this process DCs develop a mature phenotype, which allows them to present antigens captured in the periphery and to stimulate proliferation of antigen-specific T cells.

Alongside the induction of adaptive immune responses, a second major DC function is the prevention of autoimmunity. Immature DCs loaded with self-antigens in the absence of activating stimuli induce apoptosis or anergy in autoreactive T cells (reviewed in Reis e Sousa, 2006). Mature DCs can also contribute to the establishment of self-tolerance by promoting the expansion of regulatory T cells (Oldenhove et al., 2003). Thus, depending on the signals from their microenvironment, DCs can either elicit immunogenic or tolerogenic responses (reviewed in Villadangos and Schnorrer, 2007).

Since their first description in 1973 by Steinman and Cohn (Steinman and Cohn, 1973), knowledge about DC biology has increased considerably, as methods became available to generate large numbers of DCs *in vitro* (Inaba et al., 1992; Maraskovsky et al., 1996). It has become clear that the archetypical DC life cycle described in the Langerhans cell paradigm does not adequately depict the complex real situation (reviewed in Villadangos and Heath, 2005). Studies in the recent decades have revealed that DCs are a heterogenous group of cells with differences in anatomical location, life cycle, surface markers and function (reviewed in Heath and Carbone, 2009; Coquerelle and Moser, 2010; Guilliams et al., 2010).

4.1.1 DC subsets

Presumably due to the existence of differing pathogens, routes of infection and immune escape mechanisms, multiple subsets of DCs have evolved. However, DCs from all subsets share some common features, e.g., they all constitutively express the integrin CD11c. Moreover, upon maturation, they all upregulate major histocompatibility complex (MHC) II molecules on their surfaces and exhibit a characteristic dendritic morphology (reviewed in Naik, 2008; Steinman and Idoyaga, 2010).

In the past, various DC classification systems have been employed, for example immature versus mature DCs, resident versus migratory DCs, or lymphoid tissue versus non-lymphoid tissue DCs (reviewed in Steinman and Idoyaga, 2010). More recently, DCs have been rather subdivided into two major categories based on their function: classical (also known as conventional) DCs (cDCs), a category which itself comprises many subsets, and plasmacytoid DCs (pDCs). Both will be described in more detail in sections 4.1.1.1 and 4.1.1.3, respectively. New studies have revealed that even in the skin several specialized DC subsets exist (Bursch et al., 2007; Henri et al., 2009). However, these are likely to be correlates of DC subsets in other tissues (Edelson et al., 2010). In consequence, Guilliams and colleagues have proposed a simplified DC classification system, based on functional similarities of DC subsets in lymphoid and non-lymphoid tissues. This model suggests that there are at least five types of DCs: two distinct types of cDCs, Langerhans cells (LCs), pDCs and monocyte-derived inflammatory DCs (reviewed in Guilliams et al., 2010). For reasons of clarity, this work will describe only mouse but not human DC subsets in the following sections.

4.1.1.1 $CD11b^+$-like and $CD8\alpha^+$-like cDCs

In the spleen, cDCs can be subdivided on the basis of $CD8\alpha$ expression into $CD8\alpha^-CD11b^+$ and $CD8\alpha^+ CD11b^-$ DCs. The $CD8\alpha^-$ DCs are specialized in presentation of exogenous antigen via MHC II, whereas $CD8\alpha^+$ DCs are specialized in presentation of endogenous self or viral antigens via MHC I (reviewed in Merad and Manz, 2009; Pooley et al., 2001). Remarkably, $CD8\alpha^+$ DCs are also able to present exogenous antigen via MHC I, a process known as 'cross-presentation' (den Haan et al., 2000). In the LNs, in addition to these two splenic subsets, tissue-derived migratory DCs are also found (reviewed in Randolph et al., 2008). Epidermal LCs and dermal DCs (dDCs), for example, continuously enter the skin-draining LNs. Two types of migratory DCs have been linked to LN-resident DCs based on developmental and functional studies: langerin$^-$CD11b$^+$ dDcs seem to be related to CD8α$^-$CD11b$^+$ DCs (del Rio et al., 2007), whereas langerin$^+$CD103$^+$ dDCs share some characteristics with CD8α$^+$ DCs, particularly their cross-presentation ability (Edelson et al., 2010).

Thus, several cDC subsets appear highly related and therefore several authors have suggested to regroup them into the major CD11b$^+$-like and CD8α$^+$-like cDC subsets, whose main function is the priming of antigen-specific helper CD4$^+$ T cells and cytotoxic CD8$^+$ T cells, respectively (reviewed in Guilliams et al., 2010; Heath and Carbone, 2009).

4.1.1.2 Langerhans cells

DCs were initially discovered in 1868 by Paul Langerhans, who described dendritical cells in the human skin, which subsequently were named after him. However, he assumed that they were nerve cells (Langerhans, 1868). Over 100 years later, it was recognized that Langerhans cells are leukocytes and represent the DC population of the epidermis (Schuler and Steinman, 1985). LCs form a dense network in the skin and also occur in various mucosal tissues, where they build a first immunological barrier against pathogens that breach body surfaces (reviewed in Iwasaki, 2007). Their characteristic feature are 'tennis racket'-shaped Birbeck granules, which can be visualized by electron microscopy. These unique granules are induced by langerin, a C-type lectin receptor that is expressed at high levels in LCs (Kissenpfennig et al., 2005). The exact role of Birbeck granules is not known, but their component langerin presumably functions as an antiviral receptor (reviewed in van der Vlist and Geijtenbeek, 2010), for example for HIV-1 (de Witte et al., 2007). In addition, binding to a variety of pathogenic cell surface carbohydrates has been observed (Feinberg et al., 2011). Expression of langerin is not unique to LCs, as several groups have shown that a langerin-positive DC population also exists in the dermis (Bursch et al., 2007; Ginhoux et al., 2007; Poulin et al., 2007). Additionally, some LN-resident CD8α$^+$ DCs express langerin (Douillard et al., 2005; Flacher et al., 2008).

LCs differ from other DCs in multiple respects: (i) LC development is independent of FMS-like tyrosine kinase 3 ligand (Flt3L) (section 4.1.2) (ii) 10 days after birth the epidermal LC network is established due to an 'explosive' proliferation of LC precursors during the first week (Chorro et al., 2009) (iii) LCs are maintained by self-renewal or by local precursors in the absence of inflammation (Merad et al., 2002) (iv) LCs are replaced by circulating monocytes under inflammatory conditions (Ginhoux et al., 2006) (v) LC turnover is slow, with an estimated half-life of several weeks (Vishwanath et al., 2006) and (vi) LCs are radioresistant (Collin et al., 2006).

Recent studies using LC-deficient mice challenged the classical LC paradigm (reviewed in Kaplan et al., 2008). There have been conflicting results concerning the precise function of LCs, as many earlier studies on skin DCs did not distinguish between the langerin$^+$ LCs and the in 2007 discovered langerin$^+$ dDCs (Bursch et al., 2007). Novel *in vivo* findings suggest that LCs cannot cross-present antigen, whereas langerin$^+$ dDCs can (Bedoui et al., 2009). By measuring contact hypersensitivity (CHS) responses to topically applied haptens, it now has become clear that LCs can actively suppress immune responses through the production of IL-10 (Igyarto et al., 2009). Therefore, LCs appear to have an immunoregulatory role and might function to suppress responses against commensal microorganisms (reviewed in Kaplan, 2010).

4.1.1.3 pDCs

While pDCs and cDCs share common progenitors (section 4.1.2), pDCs possess distinct molecular features resembling those of lymphocytes. pDCs are also called natural interferon-producing cells, as they produce massive amounts of type I interferons (IFN-α and -β) in response to viruses (reviewed in Liu, 2005). As a consequence, natural killer (NK) cells are activated to destroy virus-infected cells (Krug et al., 2004). Thus, pDCs are functionally specialized in antiviral defense. Pathogen-detection is mediated by pDC expression of the endosomal nucleic acid-sensing Toll-like receptors (TLRs) TLR7 and TLR9, which bind to single-stranded RNA or unmethylated CpG-containing DNA, respectively (reviewed in Colonna et al., 2004). TLR signaling is required to induce pDC activation and upregulation of MHC II and costimulatory molecules on the pDC surface, which allows them to efficiently present and cross-present antigen (reviewed in Villadangos and Young, 2008). In contrast to cDCs, pDCs continue to synthesize peptide-MHC II complexes after activation, and thus the repertoire of viral antigens presented by cDCs and pDCs is thought to be largely non-overlapping (Young et al., 2008). While pDCs exhibit a secretory lymphocyte-like morphology in the steady-state, they gain a more dendritic cell-like shape with cytoplasmic protrusions upon maturation (reviewed in Soumelis and Liu, 2006). Besides their morphology, pDCs share some transcription factors with lymphocytes, such as E family proteins, which are important in lymphopoiesis. The continuous expression of the E protein E2-2 in pDCs has recently been shown to be critical for the maintenance of the pDC lineage fate, as deletion of E2-2 in mature pDCs causes spontaneous differentiation into cDC-like cells (Cisse et al., 2008; Ghosh et al., 2010).

Thus, the pDC lineage exhibits some plasticity, and the relationship of pDCs to cDCs is controversial (reviewed in Reizis, 2010).

4.1.1.4 Inflammatory DCs

Monocytes are the precursors of macrophages as well as of different subsets of DCs. They are rapidly recruited to sites of infection in response to inflammation. In the presence of granulocyte macrophage colony-stimulating factor (GM-CSF), monocytes can differentiate into $CD11c^+MHC\ II^+$ DCs (Randolph et al., 1999), that fulfill a crucial role in T cell priming (reviewed in Dominguez and Ardavin, 2010). Similarly, TiP (TNF/iNOS (inducible nitric oxide synthase)-producing) DCs are generated from monocytes upon infection with certain pathogens, e.g., *Listeria monocytogenes* (Serbina et al., 2003). Due to their capacity to produce high amounts of microbicidal TNF and iNOS, TiP DCs play a role in the clearance of bacterial infections (reviewed in Dominguez and Ardavin, 2010). Thus, depending on the kind of infection or inflammation, monocytes can differentiate into distinct monocyte-derived inflammatory DCs (reviewed in Shortman and Naik, 2006). This plasticity prompted Hume to doubt whether DCs and macrophages are actually different cell types (Hume, 2008).

4.1.2 DC development

DC lineage and terminal differentiation are controlled by the interplay of specific cytokines and transcription factors. Moreover, DC differentiation is critically influenced by miRNAs (microRNAs) (Publication I/section 7.1). DCs as well as all other leukocytes develop from BM-derived hematopoietic stem cells (reviewed in Liu and Nussenzweig, 2010). Lymphoid and myeloid lineages diverge early in hematopoiesis. While a common lymphoid progenitor (CLP) can differentiate into T, B and NK cells, a common myeloid progenitor (CMP) gives rise to monocytes, macrophages, granulocytes, megakaryocytes, erythrocytes and DCs (Akashi et al., 2000). Via CMP-derived macrophage-DC progenitors (MDPs), common DC progenitors (CDPs) can develop in a Flt3-dependent manner. These CDPs are of the phenotype $Lin^-CD115^+Flt3^+ckit^{lo}$ and can produce cDCs as well as pDCs. Downstream of CDPs are the migratory pre-DCs ($Lin^-CD11c^+MHC\ II^-SIRP\alpha^{lo}Flt3^+$), which leave the BM and after a short period of circulation in the blood give rise to lymphoid and non-lymphoid tissue DCs (reviewed in Liu and Nussenzweig, 2010). LCs are an exception to this model, as they are maintained via self-renewal and/or specialized precursor cells in the skin (section 4.1.1.2).

4.1.2.1 Cytokines

There are four major cytokines critically involved in DC development: GM-CSF, M-CSF, Flt3L and TGF-β1 (reviewed in Merad and Manz, 2009). In vivo DC homeostasis is mainly dependent on the concerted action of GM-CSF and Flt3L. While in the absence of GM-CSF cDC numbers are reduced, absence of Flt3L affects both cDC and pDC numbers, in accordance with the respective cytokine receptor expression profiles (Kingston et al., 2009). Inversely, injection of Flt3L leads to expansion of both pDCs and cDCs (Maraskovsky et al., 1996). Macrophage colony-stimulating factor (M-CSF), which is a key cytokine for macrophage development (Dai et al., 2002), also has been found to increase pDC and cDC numbers (Fancke et al., 2008). In contrast to most DC subsets, LCs develop independently of Flt3L (section 4.1.1.2). Instead, LC development requires M-CSF and autocrine transforming growth factor-β1 (TGF-β1) (Kaplan et al., 2007).

The differential effects of various cytokines have also been demonstrated in *in vitro* BM cultures, and can be exploited to generate different kinds of DCs in large amounts. Flt3L drives the differentiation of cultured BM progenitor cells into pDCs and cDCs, while GM-CSF-supplemented cultures produce DCs resembling monocyte-derived inflammatory DCs (Gilliet et al., 2002; Xu et al., 2007).

4.1.2.2 Transcription factors

Binding of cytokines to their cognate cellular receptors results in downstream signaling and the activation of transcription factors. On the one hand, global DC development requires transcription factors that are expressed at early stages of hematopoiesis, such as PU.1, Gfi1 and Ikaros (reviewed in Watowich and Liu, 2010; Wu and Liu, 2007). In deletion models, their absence impaired the development of cDCs as well as pDCs. On the other hand, there are transcription factors whose absence affects particular DC subsets only and therefore are thought to control the branching of DC precursors into distinct populations. Batf3 (basic leucine zipper transcription factor, ATF-like 3), for example, belongs to the latter category, as Batf3$^{-/-}$ mice specifically lack CD8α$^+$ DCs in spleen and LNs (Hildner et al., 2008), as well as dermal CD103$^+$ DCs (Edelson et al., 2010). Another example is E2-2, whose expression is restricted to pDCs, and which is important for pDC subset differentiation. E2-2 directly induces the expression of pDC-specific genes, such as interferon regulatory factor 7 (IRF7), which is the master regulator of IFN expression, and TLR7 and TLR9 (Cisse et al., 2008). In addition to reinforcing lineage commitment, E2-2 inhibits critical determinants of the alternative cDC fate, such as Id2 (inhibitor of DNA binding 2). Inversely, Id2 expression in cDCs inhibits E2-2 function, suggesting that these two factors play antagonistic roles in cell fate decisions (reviewed in Reizis et al., 2010).

4.1.3 DC activation and maturation

DCs in non-lymphoid tissues are in an immature state, in which they sample their environment via phagocytosis, macropinocytosis and receptor-mediated endocytosis. They are equipped with a variety of pattern-recognition receptors (PRRs), such as C-type lectin receptors (e.g., langerin and DC-SIGN) and TLRs. TLRs are a major group of receptors recognizing pathogen-associated molecular patterns (PAMPs), which comprise lipids, lipoproteins, proteins and nucleic acids, derived from various pathogens including bacteria, viruses, parasites and fungi (reviewed in Kawai and Akira, 2010). In response to receptor binding, DCs upregulate CCR7 and migrate to the draining LNs, which brings them into contact with T cells. During their migration, DCs undergo a process termed maturation, a key feature of DC biology, which is induced by the activation of transcription factors such as nuclear factor-κB (NF-κB) or activating protein-1 (AP-1) (reviewed in Kawai and Akira, 2006). Maturation involves the upregulation of several surface markers, most importantly MHC II and the costimulatory molecules CD80, CD86 and CD40 (reviewed in Reis e Sousa, 2006). Stabilization of peptide-loaded MHC II molecules on the surface of activated DCs is responsible for an enhanced antigen presentation capacity (reviewed in van Niel et al., 2008). DC activation also leads to increased survival and thus to prolonged antigen presentation in the LNs through a modulation of anti-apoptotic Bcl-2/Bcl-xL and pro-apoptotic Bax/Bak molecules (Chen et al., 2007a). Moreover, depending on the microorganism that has led to activation, DCs start to express a specific cytokine profile, which in turn polarizes naïve T cells towards T helper (T_H) $1/T_H2/$ T_H17 or regulatory T cell (T_{reg}) differentiation (reviewed in O'Shea and Paul, 2010).

4.1.4 Antigen presentation and T cell activation

DCs continually present antigen. Antigen presentation via MHC I or II requires the degradation of proteins into peptides. Peptides derived from exogenous proteins are typically loaded onto MHC II molecules and presented to helper $CD4^+$ T cells, while endogenous peptides are normally presented via the MHC I pathway to cytotoxic $CD8^+$ T cells. A unique feature of $CD8\alpha^+$-like DCs is the delivery of exogenous antigens to the MHC I pathway, known as cross-presentation (reviewed in Villadangos and Schnorrer, 2007). This mechanism is thought to be important in immunity to viruses and other intracellular pathogens. However, cross-presentation in the absence of inflammatory stimuli leads to peripheral tolerance, a mechanism known as 'cross-tolerance', which is important to prevent autoimmune disease (Luckashenak et al., 2008). The DC maturation status ('mature' is defined here as 'expressing high levels of MHCII and costimulatory molecules') can determine whether DCs have an immunogenic or tolerogenic function. Classically, mature DCs were considered to be immunogenic, whereas immature DCs were thought to have a tolerogenic function. However, it has become clear that tolerogenic DCs can also be phenotypically mature. The varying DC effector functions can be explained by their delivery of a 'signal 3' that determines T cell fates (reviewed in Reis e Sousa, 2006). Thus, besides 'signal 1' (engagement of the T cell receptor (TCR) with an appropriate MHC-peptide complex) and 'signal 2' (costimulatory signal delivered by binding of CD80 or CD86 to CD28), a third polarizing signal is necessary to direct T cell differentiation into the different subsets of effector T cells. For example, 'signal 3' might be interleukin (IL)-12, which promotes T_H1 or cytotoxic T cell (CTL) differentiation, or a Notch ligand to drive T_H2 differentiation. The concurrent presence of IL-6 and TGF-β1 leads to the differentiation of T_H17 cells. TGF-β1 only, however, induces T_{reg} differentiation (reviewed in Coquerelle and Moser, 2010).

4.1.5 DCs in clinical therapy

In clinical settings, DCs might represent a powerful tool to strengthen the body´s defense mechanisms against infectious diseases and tumors, but also to suppress unwanted autoimmune responses or allergies. A central issue in immune modulation is the identification of the pathways and factors that induce DC-mediated immunity versus tolerance, i.e. the factors guiding T_H and T_{reg} cell differentiation. For this purpose, better descriptions of the functional properties of different DC subsets are desirable. A major obstacle so far is the transfer of findings made in mouse models to the human system, as the present knowledge about human DCs is mostly derived from *in vitro* differentiated blood monocytes. However, it has been shown that certain gene signatures are evolutionarily conserved between corresponding human and mouse DC subsets (Robbins et al., 2008). For example, the transcription factor E2-2 plays a key role in both mouse and human pDC differentiation. A significant similarity in gene expression has also been revealed between mouse $CD8\alpha^+$ DCs and human BDCA3$^+$ DCs, suggesting that human BDCA3$^+$ DCs have a similar cross-presentation function as mouse $CD8\alpha^+$ DCs (Crozat et al., 2010). Indeed, it has been shown that BDCA3$^+$ DCs induce CTL responses and therefore are currently considered to be the most relevant target for DC-based therapy (Jongbloed et al., 2010). To date, only limited success has been achieved in cancer patients undergoing DC-based immunotherapy, which might be due to the patients' advanced disease states as well as the conditioning regimens in previous cancer therapies (reviewed in Tacken et al., 2007). An additional obstacle that has to be overcome is the immunosuppressive environment that is often built by established tumors, making efficient T cell activation by DCs difficult (reviewed in Melief, 2008). There are two major approaches to use DCs as adjuvants in vaccination strategies: 'DC vaccination' and 'in vivo DC targeting'. Both approaches are safe, well tolerated and present very few side effects (reviewed in Tacken et al., 2007).

4.1.5.1 DC vaccination

In DC vaccination strategies, *ex vivo* antigen-loaded DCs are adoptively transferred to patients with the aim of inducing immunity (reviewed in Schuler, 2010). DCs are generated from monocytes or $CD34^+$ precursors isolated from patient blood, and mature DCs are obtained by culturing the DCs with a cytokine mixture. Antigens that are loaded may be peptides or derived from tumor cell lysates. Another possibility is to transfect the cells with messenger RNA (mRNA) encoding for the desired peptide. The major advantage of the DC vaccination method, as compared to *in vivo* DC targeting (section 4.1.5.2), is that the DC type and maturation status can be controlled in order to ensure optimal immunogenicity. However, a tailor-made procedure for each individual is required, making it an expensive approach. Mostly, these immunization strategies have yielded very occasional success, e.g. in the treatment of melanoma (Palucka et al., 2006). Only recently, for the first time, a phase III clinical trial has provided evidence that DC vaccination has a real benefit to cancer patients. Improved overall survival has been observed in men with advanced prostate cancer, who were treated with autologous DCs that had been cultured with a fusion protein composed of GM-CSF and a tumor-specific antigen (Higano et al., 2009). DC vaccines have also yielded promising results in HIV therapy (Lu et al., 2004). After vaccination with HIV-1-pulsed autologous DCs, HIV-specific $CD4^+$ and $CD8^+$ T cells were induced and viral load was significantly decreased for at least 1 year.

4.1.5.2 In vivo DC targeting

Loading of DCs with antigens *in vivo* is another promising therapeutic approach (reviewed in Tacken et al., 2007). In contrast to *ex vivo* loading, large-scale production of vaccines is possible due to the MHC haplotype independent strategy, making it less cost-intensive and available to a large number of patients. Moreover, it provides the opportunity to deliver antigens within the natural environment to endogenous DC subsets. Distinct subsets can be targeted by the choice of specific antibodies directed against defined receptors on the DC surface. Injection of antibody-antigen conjugates can induce tolerance in mouse models, therefore a suitable maturation stimulus needs to be co-delivered to induce immunity (reviewed in Caminschi et al., 2009). According to Aarntzen et al., approaches targeting C-type lectin receptors (CLRs) are most likely to enter the clinic in the near future (Aarntzen et al., 2008). The mannose-receptor, CD205 (also known as DEC205) and DC-SIGN, which all are CLRs implicated in receptor-mediated endocytosis, are a major focus of targeting studies. However, these receptors differ in their expression patterns, their signaling pathways and the requirement for adjuvants to induce effective responses (reviewed in Tacken et al., 2007). So far, there is evidence that anti-CD205 fusion antibodies are superior in eliciting cross-presentation and thus CTL responses in humans, as compared to anti-mannose receptor or anti-DC-SIGN fusion antibodies (Bozzacco et al., 2007). However, novel promising antibodies have been generated that more selectively target mouse $CD8\alpha^+$ DCs and potentially the human equivalent. Anti-langerin and anti-Clec9A (also known as DNGR-1) have been recently shown to induce $CD8^+$ T cell responses *in vitro* comparable to anti-CD205 (Idoyaga et al., 2011). Furthermore, delivery of antigens to Clec9A induced enhanced antibody responses, as well as enhanced T cell proliferative responses, even in the absence of additional adjuvants (Caminschi et al., 2008). Targeting of Clec9A, together with an adjuvant, has successfully cured mice of a transplantable tumor (Sancho et al., 2008). These results indicate that the choice of the target receptor is critical due to its influence on the induction of adaptive responses.

4.1.5.3 Novel strategies in DC-based tumortherapy

Although some promising results have been achieved with DC-based immunotherapy especially in mouse models, so far this method cannot cure established cancer in humans. However, in concerted action with the beneficial effects of existing chemotherapy on DCs (e.g., enhanced DC activation, cross-presentation of tumor-associated antigens from apoptotic tumor cells), novel DC-based therapies provide attractive perspectives (reviewed in Melief, 2008). Moreover, functional conditioning of DCs may lead to the generation of more successful DC vaccines.

Attempts to optimize DC vaccines are currently concentrated on the enhancement of DC immunogenicity. A mature DC phenotype with high expression levels of costimulatory molecules, moreover the production of IL-12 but no production of immunosuppressive cytokines such as IL-10, is desirable. To this end, DCs transfected with mRNAs coding for advantageous molecules such as CD70, CD40L or a constantly active TLR4 have been tested (Bonehill et al., 2009). However, with regard to the immunosuppressive tumor environment, DC vaccination efficacy may be more effectively enhanced by targeting the negative arm of DC immune regulation. Therefore, the knockdown of the immunosuppressive cytokines IL-10 and TGF-β1 or inhibitory molecules such as SOCS1 (suppressor of cytokine signaling 1) by small interfering RNAs (siRNAs) appears to be a promising strategy (Huang et al., 2011).

The development of novel strategies to improve DC immunogenicity may also benefit from increasing knowledge of the role of microRNAs (miRNAs) in immune regulation. A detailed description of miRNA biology and function follows in sections 4.2 and 4.3. Holmstrøm et al. have suggested that miRNAs can be used as biomarkers to distinguish between immature and mature DCs, similar to costimulatory molecules. Levels of selected miRNAs in DCs might even correlate with DC immunogenicity and therefore allow to assess the quality of DC vaccines (Holmstrøm et al., 2010). Moreover, as miRNAs are involved in regulating DC maturation and function (Publication II and Ceppi et al., 2009; Dunand-Sauthier et al., 2011; Jurkin et al., 2010; Lu et al., 2011; Rodriguez et al., 2007), miRNAs possibly represent targets that may be exploited in DC vaccination strategies.

4.2 miRNAs

miRNAs are a class of endogenous small non-coding RNAs that function by regulating gene expression posttranscriptionally. First described in 1993, the lin-4 miRNA was initially believed to be a unique example of a so far unknown regulatory mechanism in *Caenorhabditis elegans* (Lee et al., 1993). Processed from a short hairpin precursor, lin-4 mediates regulation of a specific mRNA through partial base pairing with its 3' untranslated region (UTR). Then, in 2000, another small RNA was detected in *C. elegans*, let-7, which, similarly to lin-4, controlled the production of a protein involved in developmental timing (Reinhart et al., 2000). The finding that the let-7 sequence was perfectly conserved across many organisms, together with the discovery of the RNA interference (RNAi) mechanism in plants, led to the insight that miRNAs are an evolutionarily ancient class of regulatory molecules with broad biological functions (reviewed in Ambros, 2008). Today, hundreds of miRNAs, many of them evolutionarily conserved across species, have been identified in mammals. The database miRBASE, for example, currently lists more than thousand human miRNAs (miRBASE release 16, http://www.mirbase.org/). Confirming the crucial role of miRNAs in animal development, Bernstein et al. have demonstrated that constitutional miRNA-deficiency in mice is lethal at an early embryonic stage (Bernstein et al., 2003). When Landgraf et al. conducted a large-scale miRNA expression profiling, they could show that distinct cell types express unique miRNA profiles, with changing patterns during cellular differentiation and malignant transformation (Landgraf et al., 2007). Consistently, Lu et al. have demonstrated that tumors can be classified based on their miRNA profiles, which reflect the developmental lineage and differentiation state (Lu et al., 2005). Moreover, many studies have demonstrated a direct link between miRNA dysregulation and the development of various diseases including cancers (reviewed in Stefani and Slack, 2008).

4.2.1 miRNA biogenesis and target recognition

Production of miRNAs proceeds via multiple processing steps (reviewed in Filipowicz et al., 2008; Krol et al., 2010). miRNAs can be encoded within intergenic regions or introns. After transcription by RNA polymerase II, the resulting long primary miRNA is processed by the enzymes Drosha and DGCR8 in the nucleus into a ~70 nt long pre-miRNA. Upon export into the cytoplasm by Exportin 5, the loop structure of the pre-miRNA is cleaved by the RNase III enzyme Dicer and its cofactor

TRBP, to yield the double-stranded ~22 nt mature miRNA. Next, one of the strands (called the guide strand) is preferentially loaded into the RISC (RNA-induced silencing complex) and stably associated with Argonaute proteins, whereas the other strand (called the miRNA* passenger strand) is degraded. The mature miRNA then serves as a guide to direct the RISC to target mRNAs bearing complementary sequences in the 3'UTR. miRNA binding to the target mRNAs results in destabilization of the targets and thus silences gene expression.

Each miRNA may control the expression of hundreds of target genes (reviewed in Bartel, 2009). As pairing of miRNAs to mRNAs is mostly imperfect, target gene predictions are not straightforward. However, several target prediction programs have been developed, for example TargetScan (program accessible at http://www.targetscan.org/), which we have used for our analyses. Existing programs have a high degree of overlap, with the major rule being that there must be perfect and contiguous base pairing between the the miRNA and the mRNA at nucleotides 2-8 of the miRNA, a region which is called the miRNA 'seed' (reviewed in Bartel, 2009).

4.2.2 Regulation of miRNA expression

miRNA accumulation in the cell is dependent on the rates of miRNA transcription, processing and decay (reviewed in Kai and Pasquinelli, 2010). miRNAs are differentially expressed in different cell types. In a similar manner to protein coding genes, differential regulation of miRNAs is controlled by epigenetic mechanisms and transcription factors (reviewed in Davis and Hata, 2009). But not only is the expression of specific miRNAs subject to regulation, their stability also seems to be controlled. miRNAs are generally considered to be rather stable molecules, with half-lives ranging from hours to days (reviewed in Kai and Pasquinelli, 2010) and an average half-life of ~5 days (Gantier et al., 2011). However, several mechanisms exist that affect their half-lives, including *cis*-acting modifications (for example 3' adenylation (Katoh et al., 2009)) and *trans*-acting enzymes (for example exonucleases (Chatterjee and Grosshans, 2009)). In addition, overall miRNA levels are regulated by the availability of particular enzymes, such as Dicer (Wiesen and Tomasi, 2009). Expression of Dicer is inhibited by cellular stresses, as well as interferons, and thus miRNA levels are also regulated by disease states.

4.2.3 miRNA regulation of gene expression

Given that more than 60 % of all human protein coding genes are predicted to contain miRNA binding sites in their 3'UTR, there must be some evolutionary pressure to maintain miRNA regulation

(Friedman et al., 2008). Also, many mRNAs have multiple miRNA binding sites, such that different coexpressed miRNAs may regulate target genes coordinately or synergistically (Krek et al., 2005; Xu et al., 2010). Even though each miRNA has the potential to repress the expression of hundreds of genes, the degree of repression of an individual mRNA appears to be relatively modest. Recent studies investigating the influence of miRNAs on protein output suggest that miRNAs typically do not change the levels of individual proteins more than 2-fold, and changes rarely exceed 4-fold (Baek et al., 2008; Selbach et al., 2008).

Although classically miRNAs were thought to act rather by 'translational repression', affecting only protein levels while leaving mRNA levels unchanged, than by 'mRNA destabilization', this view now is challenged. Comparisons of data from mRNA arrays and from ribosomal profiling, which measures total protein synthesis, indicate that the predominant reason for reduced protein output is destabilization of target mRNAs (Guo et al., 2010).

The observed subtle changes in protein concentration imply that, rather than inducing dramatic changes, miRNAs fine-tune gene expression. However, the effect of many miRNAs or miRNA clusters targeting components of the same pathway can efficiently modulate cellular functions. For example, the cluster miR-144/451 tunes the expression of many genes that allow terminal erythrocyte differentiation (Rasmussen et al., 2010). Another example is miR-181a, which targets different phosphatases and thereby tunes T cell sensitivity (section 4.3.2). Moreover, slight changes in protein concentrations might also have profound physiological effects, as seen in apoptosis, where the ratio of pro- and anti-apoptotic molecules determines the cell fate (Chen et al., 2007a). Importantly, in contrast to regulation by transcription factors, miRNAs are capable of inducing rapid changes in gene expression programs, which is of particular importance when responding to environmental cues (reviewed in Davis and Hata, 2009).

4.3 miRNAs in the immune system

Due to the rapid evolvability of miRNAs and target sites, miRNA gene control seems to be perfectly qualified to adapt to ever changing host-pathogen interactions (reviewed in Xiao and Rajewsky, 2009). In a very short time of research, it has become clear that miRNAs are highly involved in the regulation of immune pathways, with more than 100 miRNAs being expressed in cells of the adaptive and innate immune system (reviewed in O'Connell et al., 2010b). In particular, miRNAs target genes involved in signaling, for example transcription factors (Asirvatham et al., 2008), and thus act as fine-tuners of immune responses (reviewed in Lodish et al., 2008). In line with their fundamental role in the immune system, dysregulation or loss of certain miRNAs can severely compromise immune functions or cause disorders like autoimmunity or cancer (reviewed in Xiao and Rajewsky, 2009).

4.3.1 miRNA regulation of hematopoietic stem cell differentiation

Throughout the hematopoietic differentiation process, selective gene silencing by miRNAs might be essential to promote lineage commitment, as the following examples indicate. Chen et al. provided early evidence that miRNA expression in hematopoietic cells is regulated and modulates lineage commitment. Specifically, they show that ectopic expression of miR-181 directs hematopoietic stem or progenitor (lineage⁻) cells towards the B cell lineage (Chen, 2004). Also miR-223 plays a role in lineage differentiation processes. It acts as a molecular switch during granulopoiesis by targeting the transcription factors Mef2c (important for myeloid progenitor differentiation) and E2F1 (a key regulator of cell cycle progression), and thus regulates granulocyte differentiation versus myeloid cell proliferation (Johnnidis et al., 2008; Pulikkan et al., 2010). The clustered miRNAs miR-221 and miR-222 are important in erythropoiesis. Downregulation of miR-221/222 during erythropoiesis relieves repression of their target, the stem cell factor receptor c-kit, and presumably in this way promotes the expansion of early erythroblasts (Felli, 2005).

4.3.2 miRNA regulation of lymphocyte development and function

Among miRNAs expressed by hematopoietic cells, miR-155 has emerged as a miRNA with major impact in T and B lymphocyte biology. miR-155 is highly expressed upon activation of B cells, T cells, macrophages and DCs, suggesting a positive role in mediating inflammatory responses (reviewed in Baltimore et al., 2008). Consistently, responses to pathogens or immunization were impaired in miR-155 knockout (ko) mice (Rodriguez et al., 2007; Thai et al., 2007). This was, at least in part, due to increased expression of the transcription factor c-maf, which leads to T_H2 polarization and enhanced production of T_H2 cytokines. Germinal center responses also were affected in miR-155 deficient mice, and B cells failed to produce high-affinity IgG1 antibodies. This defect was related to dysregulation of PU.1, a confirmed target of miR-155 (Vigorito et al., 2007). Confirming the role of miR-155 in inflammation, it has been shown that miR-155 ko mice are resistant to experimental autoimmune encephalitis (EAE), possibly due to reduced T_H17 cell formation (O'Connell et al., 2010a).

Levels of miR-181a are dynamically regulated during thymocyte development, with the highest levels being found in immature thymocytes (Li et al., 2007; Neilson et al., 2007). As miR-181a downregulates several phosphatases (e.g., SHP-2) involved in TCR signaling, the high miR-181a levels present in immature thymocytes can reduce the signaling threshold and thereby increase T cell sensitivity. Thus, miR-181a acts as a 'rheostat', which intrinsically regulates antigen sensitivity, to allow for differential responsiveness to TCR ligation during thymic selection and in the mature state (Li et al., 2007).

Various mouse models with a conditional Dicer ko in T and B cells have demonstrated the overall requirement for Dicer-dependent miRNAs for normal immune function (reviewed in O'Connell et al., 2010b). In the models studied, Dicer was deleted at different times during cellular development, depending on the promoter used to drive Dicer excision, and therefore results cannot be compared. However, some similarities can be noticed. Defective homeostastis or increased apoptosis in the absence of Dicer has been observed in NKT cells (Fedeli et al., 2009; Zhou et al., 2009b), in T_{reg} cells (Cobb et al., 2006; Zhou et al., 2008), in $CD4^+CD8^+$ $\alpha\beta$ T cell progenitors (Cobb, 2005), as well as in B cells, where the pro-apoptotic molecule Bim was found upregulated due to absence of the miR-17~92 cluster (Koralov et al., 2008). Besides regulating apoptosis, miRNAs are also likely to participate in lymphocyte lineage fate decisions, as in miRNA-deficient $CD4^+$ T cells T_H1/T_H2 differentiation is defective (Muljo, 2005) and miRNA-deficient T_{reg} cells loose their suppressive activity (Zhou et al., 2008).

4.3.3 miRNA regulation of innate inflammatory responses

TLR triggering in macrophages and DCs results in activation of a downstream signaling cascade, the end result of which is mobilization of transcription factors such as activator protein-1 (AP-1) and NF-κB, which induce expression of pro-inflammatory cytokines and chemokines (reviewed in Akira et al., 2006). Tight regulation of TLR signaling is critical to avoid excessive inflammation. Taganov et al. have proposed a model in which miRNAs play a role in the resolution phase of inflammatory processes. They have shown that in human monocytic cells several miRNAs (i.e. miR-146a, -132 and -155) are upregulated by NF-κB in response to lipopolysaccharide (LPS)-induced TLR4 triggering. In turn, miR-146a targets IL-1R-associated kinase 1 (IRAK1) and TNFR-associated factor 6 (TRAF6), which are both important molecules involved in NF-κB activation (Taganov, 2006). Thus, miR-146a acts in a negative feedback loop to dampen pro-inflammatory responses.

As mentioned above and in section 4.3.2, expression of miR-155 is also strongly induced upon activation in macrophages and DCs. Overall evidence suggests a more pro-inflammatory than inhibitory role for miR-155 in the priming of adaptive immune responses, although its role in inflammatory responses is not yet entirely clear (reviewed in O'Neill et al., 2011). On the one hand, miR-155 has been shown to promote pro-inflammatory responses by inhibition of Src homology 2 domain-containing inositol-5'-phosphatase 1 (SHIP1) and SOCS1 (O'Connell et al., 2009; Wang et al., 2010). Furthermore, DCs appear to require miR-155 for efficient T cell priming (Rodriguez et al., 2007). On the other hand, downregulation of the pro-inflammatory cytokine IL-1 and components of the IL-1 signaling pathway by miR-155 has been observed (Ceppi et al., 2009).

An anti-inflammatory role has been suggested for miR-21, which is induced in macrophages treated with LPS, the consequence of which is enhanced IL-10 production, as well as inhibition of miR-155 expression (McCoy et al., 2010; Sheedy et al., 2009).

These findings led to the integrated view of sequential induction of different miRNAs upon TLR activation to regulate the inflammatory response (reviewed in O'Neill et al., 2011). Thus, it has been suggested that first inflammation is promoted (mainly by miR-155), then in a feedback loop TLR signaling is negatively regulated (mainly by miR-146a) and finally an anti-inflammatory response is mediated by miR-21.

4.3.4 miRNAs in hematopoietic malignancies

miRNAs are often encoded at fragile sites or in genomic regions associated with cancer (Calin et al., 2004). Given the crucial role of miRNAs in proliferation and differentiation of hematopoietic cells, it is not surprising that dysregulation of miRNAs is found in some immune cell-based cancers.

As described before (sections 4.3.1 and 4.3.2), some miRNAs are involved in regulation of cell cycle progression, while others are involved in regulation of apoptosis through targeting either pro- or anti-apoptotic molecules. The cluster miR-17~92, also known as oncomiR-1, carries out pleiotropic functions during both normal development and tumorigenesis. The cluster, which comprises six miRNAs (miR-17, -18a, -19a, -19b, -20a and -92), has been reported to be highly expressed in various human cancers, including lymphomas and leukemias (reviewed in Mendell, 2008). Transgenic mice that overexpress miR-17~92 in lymphocytes developed a lymphoproliferative disease due to increased lymphocyte proliferation and showed less activation-induced cell death. This was most likely caused by miR-17~92-induced repression of the tumor suppressor PTEN as well as of the pro-apoptotic protein Bim (Xiao et al., 2008).

Similarly to miR-17~92, miR-155 can act as an oncogene. Overexpression of miR-155 has been detected in diverse tumor samples, for example in the bone marrow of patients with acute myeloid leukemia (Garzon et al., 2008). Furthermore, confirming the oncogenic function of this miRNA, mice overexpressing miR-155 developed leukemia or, depending on the system used, a myeloproliferative disorder (Costinean et al., 2006; O'Connell et al., 2008).

Also miR-15a and miR-16 have been associated with cancer. They inhibit the pro-survival protein bcl-2 and are therefore thought to function as pro-apoptotic molecules. Their loss has frequently been observed in chronic lymphocytic leukemia (Calin et al., 2002).

In conclusion, increasing data indicate that particular miRNAs can function as oncogenes or tumor suppressors. Therefore, their dysregulation can contribute to tumor development, implying that therapeutic targeting of miRNAs might represent a reasonable and potent strategy in cancer treatment (reviewed in Garzon et al., 2010).

5 Publication I (Kuipers et al., Mol Immunol 2010)

Differentially expressed micrornas regulate plasmacytoid vs. conventional dendritic cell development

Harmjan Kuipers[*], Frauke M. Schnorfeil[*] and Thomas Brocker

Molecular Immunology. 2010 Nov-Dec;48(1-3):333-340[1]

[*] equal contribution

---

[1] Reprinted from Molecular Immunology, 48(1-3); Kuipers, H., Schnorfeil, F.M. and Brocker, T. Differentially expressed microRNAs regulate plasmacytoid vs. conventional dendritic cell development, 333-340; Copyright 2010, with permission from Elsevier.

Molecular Immunology 48 (2010) 333–340

Contents lists available at ScienceDirect

Molecular Immunology

journal homepage: www.elsevier.com/locate/molimm

Differentially expressed microRNAs regulate plasmacytoid vs. conventional dendritic cell development

Harmjan Kuipers [1,2], Frauke M. Schnorfeil [2], Thomas Brocker*

Institute for Immunology, Ludwig-Maximilian-University, Goethestrasse 31, 80336 Munich, Germany

ARTICLE INFO

Article history:
Received 7 May 2010
Received in revised form 8 July 2010
Accepted 8 July 2010
Available online 6 September 2010

Keywords:
microRNAs
Dendritic cells
miR-221
miR-222

ABSTRACT

microRNAs have emerged as a novel layer of regulation of cellular development and function, including cells of the immune system. microRNA expression profiles and function of several microRNAs have been elucidated in granulocyte macrophage colony-stimulating factor derived dendritic cells (GM-CSF DC). In this study we determined the microRNA expression profile from plasmacytoid DC (pDC) and conventional DC (cDC) generated in murine FMS-related tyrosine kinase 3 ligand (Flt3L) bone marrow culture. We observed distinct microRNA expression signatures in these two different DC subsets and found that pDC were closer related to CD4$^+$ T cells than to cDC. Expression of a selected subset of microRNAs was also compared between cDC and GM-CSF DC. Furthermore, we show that inhibition of two differentially expressed microRNAs, miR-221 and miR-222, during differentiation resulted in skewed pDC/cDC ratios. Among the confirmed or potential targets for miR-221 and miR-222 are c-Kit, p27^{kip1} and E2-2. While c-Kit is expressed by DC progenitors and p27^{kip1} is a cell cycle regulator, pDC and cDC phenotype specifically regulate pDC development. Our data demonstrate that microRNAs can influence Flt3-driven DC differentiation.

© 2010 Elsevier Ltd. All rights reserved.

1. Introduction

A novel layer of gene regulation that has a profound influence on cell differentiation and function has become visible with the discovery of non-coding RNA molecules of an average of 22 nucleotides (nt) length, so-called microRNAs (miRNAs[3]) (Bushati and Cohen, 2007). miRNAs act in a post-transcriptional fashion, binding to the 3′-untranslated region (3′-UTR) of messenger RNA (mRNA) in a sequence-specific manner, resulting in mRNA degradation or inhibition of translation (Filipowicz et al., 2008). miRNAs also regulate many aspects of the immune system. Distinct miRNA expression profiles of immunological cell types or differentiation states have been determined (Landgraf et al., 2007; Monticelli et al., 2005; Neilson et al., 2007; Wu et al., 2007) and miRNAs are involved in regulation of several innate and acquired immunological processes (Lindsay, 2008). Dendritic cells (DC) are cells from hematopoietic origin that are located in tissue and lymphoid organs and are specialized in antigen presentation (Shortman and Liu, 2002). DC differentiate from a recently identified precursor into

---

* Corresponding author. Tel.: +49 89 218075674; fax: +49 89 21809975674.
E-mail address: tbrocker@med.uni-muenchen.de (T. Brocker).
[1] Present address: Crucell Holland BV, Leiden, The Netherlands.
[2] These authors contributed equally.
[3] *Abbreviations:* miR, miRNA, microRNA; DC, dendritic cell; pDC, plasmacytoid dendritic cell; cDC, conventional dendritic cell; BM, bone marrow; SCF, stem cell factor.

0161-5890/$ – see front matter © 2010 Elsevier Ltd. All rights reserved.
doi:10.1016/j.molimm.2010.07.007

a heterogeneous population of cells that exhibit a degree of specialization in the immunological functions they perform (Naik et al., 2005; Onai et al., 2007; Reis e Sousa, 2006; Shortman and Naik, 2007). In steady-state conditions, two major subsets of DC can be distinguished based on expression of defined surface markers, conventional DC (cDC) and plasmacytoid DC (pDC). These DC subtypes can be generated *in vitro* from murine bone marrow cultures supplemented with the cytokine FMS-related tyrosine kinase 3 ligand (Flt3L) and resemble the *in vivo* pDC and cDC phenotype relatively well (Brasel et al., 2000; Brawand et al., 2002; Gilliet et al., 2002; Naik et al., 2005, 2007). The cytokine granulocyte macrophage colony-stimulating factor (GM-CSF) is also able to induce DC differentiation from bone marrow progenitors (GM-CSF DC). These cultures yield a homogenous DC population that bears some similarity with Flt3L-derived cDC based on cell surface markers, but differs in function (Weigel et al., 2002; Xu et al., 2007) and is thought to be generated from different precursors under inflammatory conditions *in vivo* (Naik et al., 2007; Shortman and Naik, 2007). miRNA expression profiles of GM-CSF DC under steady-state and inflammatory conditions have been determined (Landgraf et al., 2007) and functional roles, such as pathogen binding and CD4$^+$ T cell proliferation capacity, have been assigned to some miRNAs in GM-CSF-derived DC (Martinez-Nunez et al., 2009; Rodriguez et al., 2007). In addition, miR-21 and miR-34a have been shown to regulate GM-CSF-driven DC differentiation (Hashimi et al., 2009).

The miRNA expression levels of different Flt3L-derived DC subsets have not been established yet. In this study, we analyzed the miRNA expression profile from pDC and cDC populations using miRNA microarrays and the expression pattern of selected miRNAs was compared between cDC and GM-CSF DC using quantitative real-time PCR. A subset of differentially expressed miRNAs was identified. Two highly related miRNAs, miR-221 and miR-222, that were overexpressed in cDC were studied into more detail. *Kit*, which is expressed by DC progenitors (Naik et al., 2007; Onai et al., 2007), is an experimentally verified target for these miRNAs. In addition, E2-2 (*tcf4*), a master transcriptional regulator of pDC development (Cisse et al., 2008) contains hypothetical target sequences for miR-221 and miR-222. Inhibition of these miRNAs during DC *in vitro* Flt3L-mediated differentiation revealed a shift from cDC to pDC development.

2. Materials and methods

2.1. Mice

C57BL/6J mice were bred and maintained in a conventional facility at the Institute of Immunology (Munich, Germany) and used according to protocols approved by the local animal ethics committee.

2.2. Dendritic cell cultures

Dendritic cells were generated from bone marrow (BM) in the presence of Flt3L as described (Brasel et al., 2000), with some modifications. BM was extracted from femur and tibiae and erythrocytes lysed (R&D kit). Cells were cultured at 1×10^6 cells/ml in DC tissue culture medium (DC-TCM: RPMI, 10% FCS, Pen-Strep) in the presence of 20 ng/ml recombinant murine Flt3L (R&D) and 50 ng/ml recombinant murine stem cell factor (SCF) (Peprotech). At day 4, cells were collected, replated at 1×10^6 cells/ml in the presence of recombinant murine Flt3L. At day 8, an equal volume of DC-TCM supplemented with Flt3L was added and cells were harvested at day 10. To obtain GM-CSF DC, BM cells were grown in DC-TCM supplemented with 3% supernatant of the X-63-GM-CSF cell line (Stockinger et al., 1996) following the protocol described by Lutz et al. (1999).

2.3. Transfection of miRNA inhibitors

BM cultures containing SCF and Flt3L were started as described above. At day 2, cells were harvested and replated at 4×10^5 in 6-well plates. Pre-designed (based on miRBase v8.1 sequences) and FITC-labeled miRCURY LNA miRNA knockdown probes (Exiqon) against mmu-miR-221 (5'-aaacccagcagacaatgtagct-3'), mmu-miR-222 (5'-gagacccagtagccagatgtagct-3') and a control sequence (5'-gtgtaacacgtctatacgccca-3'; bearing no sequence homology to any known miRNA or mRNA in mouse), were mixed with Lipofectamine RNAiMAX (Invitrogen) according to manufacturer's instructions and added to cells to a final concentration of 250 nM. Twenty-four hours later, the FITC-positive cells were enriched by FACS-sorting and recultured with Flt3L. Cells were harvested at day 7 to determine the total numbers of DC and subset frequency by FACS.

2.4. Flow cytometry

Anti-FcγRII/III antibody was included in all stainings to reduce non-specific antibody binding. To sort pDC and cDC from Flt3L-derived DC cultures, cells were stained with DAPI, 120G8-FITC anti-CD11c-FITC, anti-B220-PE and anti-CD11b-APC. The DC subset frequency after transfection of miR-221 and miR-222 inhibitors was analyzed with a staining panel consisting of DAPI, anti-B220-PE, anti-CD11c-PE-Cy5.5 and CD11b-APC-Cy7. All antibodies were purchased from eBioscience or BD Biosciences unless otherwise mentioned. For all experiment events were acquired on FACSCanto2 or FACSAria flow cytometers (BD Biosciences). Flow cytometric data was analyzed using FlowJo software (Treestar).

2.5. Magnetic bead-mediated cell sorting

To separate DC subsets, pDC were first enriched using magnetic bead isolation based on anti-CD11b staining (Miltenyi Biotec MACS system), according to manufacturer's instructions. CD4$^+$ T cells were isolated from spleen and lymph nodes from naïve C57BL/6J mice using a negative selection kit (Miltenyi Biotec) following the manufacturer's protocol. Purity of CD4$^+$ T cells was around 95% of alive cells, as determined by flow cytometry.

2.6. RT-PCR

To quantify gene expression of selected genes in different DC populations, total RNA was extracted using the RNeasy kit (Qiagen), following manufacturer's instructions. RNA was reverse transcribed using SuperScript III First-Strand Synthesis System using random hexamers (Invitrogen). Quantitative PCR reactions were run in a Lightcycler PCR machine (Roche) using the LightCycler FastStart DNA Master^{PLUS} SYBR Green I kit (Roche) according to manufacturer's instructions. Primer sequences were obtained from Primerbank (Wang and Seed, 2003) and are available upon request. Expression levels were normalized to ubiquitin c and relative expression calculated using the $\Delta\Delta C_T$ method.

For quantitative real-time RT-PCR of mature miRNAs, total RNA including the low molecular weight fraction was isolated using the miRNeasy kit according to manufacturer's instructions (Qiagen). Gene-specific reverse transcription was performed for each miRNA using the Taqman MicroRNA Reverse Transcription kit and gene-specific primers (Applied Biosystems) following the manufacturer's protocol. qPCR reactions were run in a Lightcycler PCR machine (Roche) using Lightcycler Taqman Master kit (Roche) in combination with miRNA-specific Taqman miRNA assays (Applied Biosystems) according to manufacturer's instructions. miR-223 expression was determined with a primer/probe set based on miRBase 8.1 miR-223 sequence. Expression levels were normalized to small RNAs snoRNA202 or RNU19, and relative expression calculated using the $\Delta\Delta C_T$ method.

2.7. miRNA arrays

Total RNA, including the fraction of small RNA that contains miRNAs, was isolated from samples using the miRNeasy kit (Qiagen). Analysis of RNA integrity, labeling and hybridization of the samples was outsourced to Exiqon (Vedbaek, Denmark). Briefly, the quality of the total RNA was verified by an Agilent 2100 Bioanalyzer profile. A mixture of equal amounts of total RNA from all samples was used as reference sample. 2 μg total RNA from sample and reference pool were labeled with Hy3 and Hy5 fluorescent dye respectively, using the miRCURY LNA Array labeling kit (Exiqon, Denmark). The Hy3-labeled samples and a Hy5-labeled reference pool sample were mixed pair-wise and hybridized to the mercury LNA array version 8.1 (Exiqon) which contains capture probes targeting all miRNAs from all organisms annotated in miRBase version 8.1 (Griffiths-Jones et al., 2008). MiRNA annotation was updated in terms of changes in nomenclature and removal of obsolete miRNAs to reflect changes between miRBase version 8.1 and version 12.0. Synthetic miRNAs hybridizing to complementary oligos on the arrays were spiked-in to control for labeling and hybridization as well as data normalization procedures. The hybridization was performed using a Tecan HS4800 hybridization station (Tecan, Austria) and the slides were scanned by a ScanArray 4000 XL scanner (PackardBiochip Technologies, USA). Image analysis was carried out using the ImaGene 6.1.0 software (BioDiscovery, Inc., USA).

2.8. miRNA arrays data analysis

Micro array data were analyzed using the Bioconductor opensource bioinformatics software project (Gentleman et al., 2004). Briefly, image analysis output files were loaded into the software package 'limma' (Smyth, 2004). Both imperfect spots, empty spots and probes reported to be underperforming (Exiqon, personal communication) were excluded from further analysis. Background was subtracted using the 'normexp' algorithm, followed by global loess normalization. Array quality control was performed using the software package 'ArrayQualityMetrics'. For unsupervised hierarchical clustering, the median log(Hy3/Hy5) ratios of replicate spots per miRNA were calculated for every sample and standard deviation across samples was computed. The expression data from the 50 miRNAs exhibiting the highest standard deviation across samples were used to generate dendrograms, using Pearson's correlation and complete linkage analysis to compute the distance matrix. Significance of clustering was analyzed with the Pvclust package (Suzuki and Shimodaira, 2006). To detect genes differentially expressed between pDC and cDC, a linear model was fitted for every gene across arrays and gene expression differences were quantified using a moderated t-test statistic, which is based on an empirical Bayes approach (Smyth, 2004). p-Values were adjusted for multiple testing using Benjamini–Hochberg's method as incorporated in the software.

2.9. Statistical analysis

Statistical differences between the experimental groups were determined by Student's two-tailed t-test. Probabilities < 0.05 were considered to be significant.

3. Results

3.1. Purification and characterization of bone marrow-derived DC subsets

To obtain sufficient numbers of pDC and cDC to be able to analyze miRNA expression, we resorted to in vitro culturing of murine bone marrow cells in the presence of Flt3L. Populations were separated by FACS based on staining of CD11c, B220 antigen and CD11b (Fig. 1A). CD11c$^+$B220$^+$CD11b^{low} pDC identified with this staining also expressed 120G8, another pDC-discriminatory marker, confirming their pDC identity (Asselin-Paturel et al., 2003). In addition, the phenotype of sorted cell populations was verified by quantitative PCR for genes known to be differentially expressed between DC subsets (Naik et al., 2005). Expression of Siglech, Ccr9, Tcf4 and Ly6c was reduced or absent in cDC, while the cDC-specific TLR3 receptor was predominantly expressed in cDC, demonstrating that pure DC populations were obtained (Fig. 1B).

3.2. miRNA expression profiling of DC subsets

To determine the miRNA expression profile from DC subsets, RNA from sorted pDC and cDC populations was hybridized to miRNA microarrays. miRNA expression levels from peripheral CD4$^+$ T cells were also analyzed, as the miRNA expression pattern from this cell population is relatively well known in comparison to other hematopoietic cell types (Cobb et al., 2006; Monticelli et al., 2005; Muljo et al., 2005; Neilson et al., 2007). In concordance with reports from the literature, several miRNAs were found to be upregulated in T cells, in particular miR-150 (Cobb et al., 2006; Liston et al., 2008; Monticelli et al., 2005; Muljo et al., 2005) and miR-15b (Neilson et al., 2007; Zhou et al., 2008) (Fig. 2A). The relationship between the different cell lineages was further analyzed using hierarchical clustering of the relative miRNA expression levels. A distance matrix was calculated using Pearson correlation and multiscale bootstrapping was used to assess the significance of the clusters found. All biological replicates clustered together as expected, but it was found that pDC are closer related to CD4$^+$ T cells than to cDC (Fig. 2B).

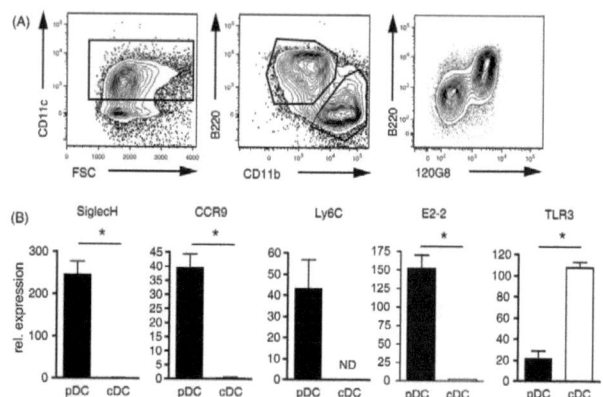

Fig. 1. Purification and characterization of bone marrow-derived DC. Bone marrow cells were differentiated towards DC in the presence of Flt3L and SCF. (A) DC cultures were stained with indicated cell surface markers and sorted into pDC (CD11c$^+$, B220$^+$, CD11b^{int}/120G8$^+$) and cDC (CD11c$^+$, B220$^{-/int}$, CD11b$^+$/120G8$^-$) subpopulations based on gates specified on plots. (B) Quantitative gene expression levels of representative genes from sorted cDC and pDC populations (n = 3–6 per group). Results are expressed in arbitrary units as mean ± SEM. *: $p < 0.05$; ND: not detected.

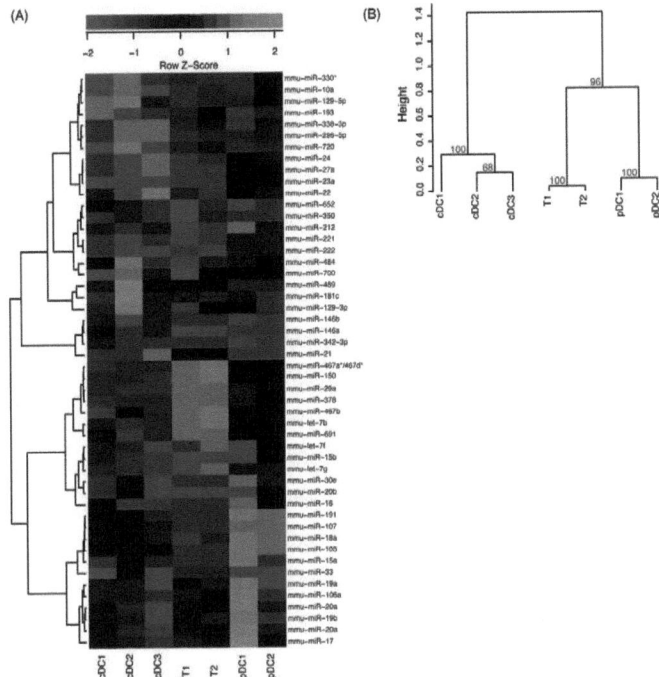

Fig. 2. miRNA expression profile from DC subsets and CD4+ T cells. (A) False-color plot representing relative expression of miRNAs from 3 cDC samples (cDC1-3), 2 pDC samples (pDC1-2) and 2 CD4+ T cell samples (T1-2). One pDC sample was a pool from 5 independent FACS sorts. The color scale shown at the top illustrates the relative expression level of the indicated miRNA across all samples: red denotes an expression below the mean and green denotes an expression higher than the mean. (B) Relationship of cell types as determined by Pearson correlation of miRNA expression profiles. Values at branch points denote multiscale bootstrapping significance percentages. (For interpretation of the references to color in this figure legend, the reader is referred to the web version of this article.)

Next, we concentrated on differences in miRNA expression between DC subsets. Differential expressed miRNAs were identified using linear modeling analysis in combination with empirical Bayes smoothing of the t-test statistic, which has been shown to be able to detect small differences in expression levels when a small number of arrays is used (Smyth, 2004). Table 1 contains the 25 miRNAs most significantly differentially expressed between cDC and pDC, ordered by the cDC/pDC expression ratio. Many miRNAs that are upregulated in cDC have also been identified in small RNA cloning studies of in vitro GM-CSF-supplemented DC cultures (Landgraf et al., 2007), in particular miR-21, miR-22 and miR-223, implying an overlap between these DC subsets in terms of miRNA expression. miR-223 has also been identified in cDNA libraries obtained from NOD-derived CD11c-positive dendritic cells (Fukao et al., 2007). Another interesting observation is the simultaneous overexpression of miR-23a, -27a and -24 in cDC, which are known to be transcribed as a single transcript (Lee et al., 2004).

3.3. miR-221 and miR-222 influence DC subset development

To determine whether changes in expression levels of particular miRNAs had functional consequences for DC development, we inhibited the function of two miRNAs, miR-221 and -222, during Flt3L-mediated DC differentiation. miR-222 was chosen from the list of differentially regulated targets for the finding that Kit mRNA is a confirmed target (Felli et al., 2005; Poliseno et al., 2006). Kit, encoding c-Kit protein, is expressed by a recently identified DC precursor capable of generating pDC and cDC (Naik et al., 2007; Onai et al., 2007). miR-221 was included due to its sequence homology to miR-222, with an identical seed region (nucleotides 2-7, Fig. 3A), and was also strongly expressed in cDC. According to the target prediction program TargetScan (Lewis et al., 2005) (www.targetscan.org), both, miR-221 and miR-222 hypothetically target tcf4, the gene encoding E2-2 transcription factor, which is a master regulator for pDC development (Cisse et al., 2008). Quantitative RT-PCR confirmed the differential expression of miR-221 and miR-222 in between DC subsets (Fig. 3B). To specifically inhibit miR-221 and -222 function, bone marrow cells differentiated to DC in the presence of Flt3L were transfected with FITC-labeled antisense oligos complementary to mature miR-221 and -222. Twenty-four hours later, FITC-positive cells were enriched by FAC sorting (average of 14.65% FITC+, 10.34-18.96 95% CI pre-sort, average of 73.69% FITC+, 68.81-78.56 95% CI post-sort; Fig. 3C) and re-cultured in the presence of Flt3L for 5 days. At day 7, the yield and subset development of the DC culture was analyzed. FITC signal could not be detected at this point (data not shown), and we

Table 1
Differential microRNA expression between cDC and pDC[a].

Name	p-Value	cDC/pDC expression ratio
mmu-miR-21	6.28^{-10}	7.28
mmu-miR-720	9.96^{-10}	5.65
mmu-miR-342-3p	1.91^{-08}	3.94
mmu-miR-296-5p	1.01^{-03}	3.80
mmu-miR-223	6.28^{-10}	3.15
mmu-miR-22	6.99^{-06}	3.07
mmu-miR-146a	9.63^{-08}	2.98
mmu-miR-222	1.31^{-05}	2.53
mmu-miR-23a	6.99^{-06}	1.94
mmu-miR-27a	2.23^{-04}	1.83
mmu-miR-24	4.85^{-04}	1.77
mmu-miR-221	8.14^{-04}	1.71
mmu-miR-23b	1.27^{-03}	1.67
mmu-miR-29a	2.23^{-04}	0.68
mmu-let-7i	8.14^{-04}	0.61
mmu-let-7a	1.95^{-04}	0.59
mmu-miR-103	2.02^{-07}	0.51
mmu-let-7g	4.38^{-04}	0.50
mmu-miR-20a	3.42^{-04}	0.46
mmu-miR-191	6.92^{-06}	0.45
mmu-miR-18a	1.66^{-05}	0.42
mmu-miR-107	2.99^{-08}	0.40
mmu-miR-15b	5.87^{-06}	0.38
mmu-miR-106a	3.02^{-04}	0.36
mmu-miR-19b	6.69^{-04}	0.33

[a] cDC and pDC show differential expression of miRNAs as determined by miRNA microarray. Total RNA from 3 cDC samples and 2 pDC samples (pDC samples pooled from 5 independent FACS sorts each) was subjected to miRNA analysis; for the exact array procedure and analysis see Section 2. The table lists the 25 most significantly (Benjamini–Hochberg's adjusted p-value) differentially expressed miRNAs, ordered by the cDC/pDC expression ratio.

did not detect differences in frequencies of alive or CD11c-positive cells (data not shown). However, when miR-221 and-222 were inhibited early in DC development a consistent decrease in cDC frequency accompanied by a small increase in pDC frequency could be observed compared to cells transfected with a control antisense oligo, resulting in an increase in pDC/cDC frequency ratio (Fig. 3D). These findings suggest that both miR-221 and miR-222 might regulate development and lineage fates of DC.

3.4. Comparison of miRNA expression profile in cDC and GM-CSF DC

Both Flt3L and GM-CSF induced differentiation of bone marrow progenitors result in DC populations in vitro. To analyze the relationship between these cDC populations developed under different cytokine stimuli at the miRNA level, we quantified the expression of several miRNAs based on the previously determined expression profile and published results (Landgraf et al., 2007). Several miRNAs are equally expressed in DC populations derived from both culture systems, including miR-223 which expression is restricted to the myeloid compartment of the murine hematopoietic system (Fukao et al., 2007; Landgraf et al., 2007). Another subset of quantified miRNAs was higher expressed in cDC compared to GM-CSF DC, among them miR-221 and miR-222. The increased expression of these two miRNAs in cDC in comparison to both pDC and GM-CSF DC might indicate an unique role of these miRNAs in cDC differentiation or function. miR-146b was the single miRNA we identified that exhibited increased expression in GM-CSF DC, in concordance with its suggested role of providing negative feedback under inflammatory conditions (Taganov et al., 2006) (Fig. 4).

4. Discussion

The DC population can be further subdivided into subpopulations, with a large distinction between pDC and cDC subsets based

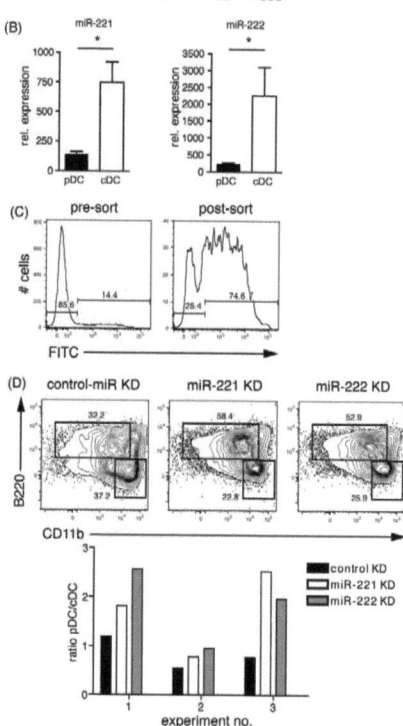

Fig. 3. miR-221 and miR-222 influence DC subset development. (A) miR-221 and miR-222 sequence alignment. (B) miR-221 and miR-222 transcript levels determined by quantitative RT-PCR from sorted cDC (n = 4) and pDC (n = 6) subset samples. Results are expressed in arbitrary units as mean ± SEM. *: p < 0.05. (C) FACS enrichment of bone marrow cells transfected with FITC-labeled miRNA inhibitors. The percentages of cells in each gate among DAPI⁻ cells are indicated. (D) B220 and CD11b expression of DC cultures of bone marrow cells transfected with indicated miRNA inhibitors (knockdown; KD). Plots shown are gated on alive, CD11c⁺ cells and representative for one out of three independent experiments. The graph depicts the ratio of pDC (B220⁺, CD11b^{int}) to cDC (B220$^{-/int}$, CD11b⁺) frequencies in cultures transfected with indicated miRNA inhibitors in the three independent experiments shown.

on several properties such as hematopoietic origin, life cycle and functional properties (Shortman and Liu, 2002; Shortman and Naik, 2007). The cDC subset can be further subdivided in two populations based on the presence or absence of the cell surface marker CD8α (Shortman and Naik, 2007). In this study we resorted to the use of a Flt3L-based in vitro cell culturing system to obtain sufficient numbers of pDC and cDC to be able to perform microarray-based analysis of miRNA expression. CD8α is not expressed by cDC in Flt3L-based culturing systems, but a functionally and phenotypically equivalent population can be identified by expression of CD24 and CD11b (Naik et al., 2005). To obtain the cDC population we gated on CD11b^{hi} cells, which are the in vitro counterparts of CD8α−

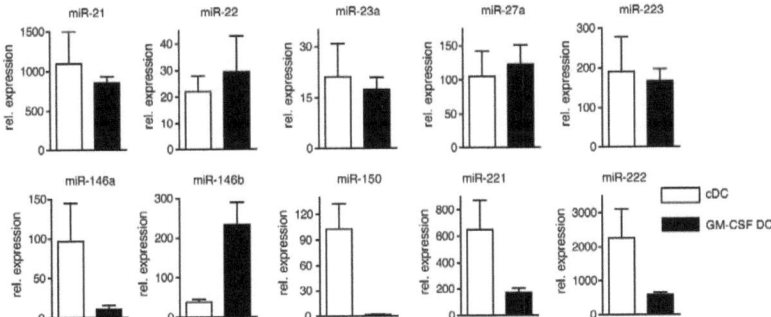

Fig. 4. Relative expression levels of miRNAs in cDC and GM-CSF DC. Bone marrow cells were cultured in the presence of Flt3L or GM-CSF. RNA was isolated from FACS sorted cDC populations (n = 4) and GM-CSF DC (n = 2). miRNA transcript levels were determined by quantitative RT-PCR. Results are expressed in arbitrary units as mean ± SEM.

negative cDC (Naik et al., 2005), although this sorted population also contained the in vitro CD8α+ cDC equivalents, as it is the only DC population to express TLR3 (Edwards et al., 2003). It would therefore be of interest to further purify the cDC subpopulation on the basis of CD24 marker expression to be able to determine the miRNA expression pattern of cDC subpopulations.

miRNA expression profiling of pDC, cDC and CD4+ T cells revealed that pDC and CD4+ T cells cluster together in terms of miRNA expression pattern. This is in contrast to a large study comparing gene expression profiles of several DC subsets in conjunction with other cell types of immunological origin, which found that DC subsets cluster together and are distinct from CD4+ T cells (Robbins et al., 2008). It remains to be determined whether this difference in lineage relationships observed reflects a true biological difference between gene expression and miRNA expression or represents an artifact due to expression profiling of a restricted number of miRNAs. Nevertheless, expression by pDC of genes having well defined function in lymphocyte populations is in support of our miRNA expression data, as pDC express the Rag gene and contain IgH D-J rearrangements (Corcoran et al., 2003; Pelayo et al., 2005; Shigematsu et al., 2004). It would also be of interest to compare the miRNA expression profile of naïve B cells to pDC, as pDC also express many B cell lineage related genes (Pelayo et al., 2005). It must also be noted that the pDC lineage is not a homogenous population by itself, as only a subset of pDC with distinct functional properties exhibit expression of Rag (Corcoran et al., 2003; Pelayo et al., 2005). Subdividing these pDC populations on the basis of Rag expression might therefore results in segregation of pDC miRNA expression profiles.

miR-21 is the most significantly differentially expressed miRNA between cDC and pDC. It is also highly expressed in GM-CSF DC, as shown in this study and Hashimi et al. (2009) and Landgraf et al. (2007). A recent study identified miR-21 to be involved in human monocyte-derived DC differentiation. miR-21 expression during monocyte differentiation in the presence of GM-CSF decreases levels of WNT1 which inhibits DC differentiation (Hashimi et al., 2009). Whether miR-21 also promotes Flt3L-derived cDC development is currently being studied. The strong expression of miR-21 in the cDC subset might also be a consequence of signal transducer and activator of transcription-3 (STAT3) expression, as two highly conserved binding sites for STAT3 have recently been identified in the region upstream of the miR-21 transcription start site, responsible for STAT3-mediated miR-21 expression (Loffler et al., 2007). STAT3 is required for Flt3L-mediated differentiation from hematopoietic stem cells into both pDC and cDC subsets (Esashi et al., 2008; Laouar et al., 2003), so STAT3-driven miR-21 expression cannot solely account for the difference in miR-21 levels between pDC and cDC.

We focused on the function of miR-221 and miR-222 in DC subset development as Kit, encoding for c-Kit protein, is expressed on a recently identified DC progenitor (Naik et al., 2007; Onai et al., 2007). We could not detect any difference in c-Kit levels on mature DC subsets by flow cytometry after inhibition of these miRNAs (data not shown), but the small but consistent decrease of the number of cDC and increase of pDC does suggest that these miRs influence DC subset differentiation. Another experimentally verified target for these miRNAs is p27[kip1], a cell cycle inhibitor that is suppressed in some human cell cancer lines by miR-221-222 resulting in continuous proliferation (le Sage et al., 2007). Recently, the murine ortholog of p27[kip1] was also found to be a target of the miR-221-222 cluster (Mayoral et al., 2009). In mast cells, overexpression of this miRNA cluster leads to cell cycle arrest and perturbed morphology (Mayoral et al., 2009). As cDC also express p27[kip1] (data not shown), the reported higher proliferation rate of cDC relative to pDC might therefore be a consequence of miR-221-222-mediated repression of p27[kip1] in the cDC subset. A third potential target for miR-221 and miR-222 is Tcf4, encoding E2-2, the master transcription factor for pDC (Cisse et al., 2008). This factor is preferentially expressed in pDC (Cisse et al., 2008 and Fig. 1B). In E2-2-deficient mice pDC development stops at an immature precursor stage leading to complete lack of mature B220+CD11b− pDC (Cisse et al., 2008). Given the fact that our TargetScan analysis identified this key molecule for pDCs as potential target for the above-mentioned miRs, a release of E2-2 suppression could explain why pDC ratios increase in anti-miR-221 or -222 treated cultures. In addition, cDC express high levels of miR-221 and miR-222 (Fig. 3B), which might lead to a decrease of E2-2 expression beyond levels of detection (Fig. 1B). However, in order to unambiguously identify the target(s) of miR-221 and miR-222 responsible for pDC/cDC development, specific miR-deficient mice will have to be generated.

Analysis of relationship between different DC subset based on gene expression patterns has shown that GM-CSF DC cluster together with monocytes and macrophages and are distinct from cDC and pDC (Robbins et al., 2008). However, despite the myeloid gene signature, some cDC-specific genes were strongly expressed in GM-CSF DC. Comparison of miRNA expression patterns using quantitative PCR of the cDC population present in Flt3L-bone marrow cultures with DC obtained after differentiation in the presence of

GM-CSF revealed that some miRNAs are similarly expressed in both DC subsets, including miR-223. This miRNA has been shown to be specifically expressed in the myeloid compartment of the murine hematopoietic system (Chen et al., 2004) and was also found to be highly expressed in GM-CSF DC (Hashimi et al., 2009; Landgraf et al., 2007). As it is lower expressed in pDC, miR-223 might therefore play a role in reinforcing phenotypic characteristics shared between cDC and GM-CSF DC. It would be of interest to analyze the phenotypes of these DC subsets from bone marrow of miR-223 deficient animals (Johnnidis et al., 2008). Interestingly, among the miRNAs differentially expressed between cDC and GM-CSF DC we identified contrasting differential expression of miR-146a and miR-146b. Both miRNAs target the mRNAs of the signaling molecules TRAF6 and IRAK1 and are postulated to constitute a negative feedback loop to fine-tune TLR signaling pathways after TLR ligand stimulation (Taganov et al., 2006). We found a higher expression of miR-146b but a lower expression of miR-146a in GM-CSF DC compared to cDC. This suggests a miR-146b-mediated decrease in TLR signaling sensitivity in GM-CSF DC, which might be explained by the pro-inflammatory nature of GM-CSF. The discrepancy between expression profiles of these two miR-146 family members needs further investigation.

In the present study we have analyzed miRNA expression profiles of different DC populations, identifying subsets of differentially expressed miRNAs. We also demonstrated that modulation of miR-221 and miR-222 expression influences DC subset differentiation. These findings warrant further investigation into the role of miRNAs in DC differentiation and function and might ultimately lead to new ways to modulate DC differentiation and function for therapeutic purposes.

Acknowledgements

This work was supported by grant KU 2513/1-1 of the Deutsche Forschungsgemeinschaft (TB) and a Rubicon grant 825.06.013 from the Dutch Scientific Organization (HK).
We would like to thank A. Bol and W. Mertl for excellent animal care and Jörg Mages from the Institute for Medical Microbiology, Immunology and Hygiene of the Technical University Munich for assistance with miRNA array data analysis.
Microarray data was deposited in the Array Express database (http://www.ebi.ac.uk/microarray-as/ae/) under accession number E-MTAB-186.
The authors have no conflicting financial interests.

References

Asselin-Paturel, C., Brizard, G., Pin, J.J., Briere, F., Trinchieri, G., 2003. Mouse strain differences in plasmacytoid dendritic cell frequency and function revealed by a novel monoclonal antibody. J. Immunol. 171, 6466–6477.
Brasel, K., De Smedt, T., Smith, J.L., Maliszewski, C.R., 2000. Generation of murine dendritic cells from flt3-ligand-supplemented bone marrow cultures. Blood 96, 3029–3039.
Brawand, P., Fitzpatrick, D.R., Greenfield, B.W., Brasel, K., Maliszewski, C.R., De Smedt, T., 2002. Murine plasmacytoid pre-dendritic cells generated from Flt3 ligand-supplemented bone marrow cultures are immature APCs. J. Immunol. 169, 6711–6719.
Bushati, N., Cohen, S.M., 2007. microRNA functions. Annu. Rev. Cell Dev. Biol. 23, 175–205.
Chen, C.Z., Li, L., Lodish, H.F., Bartel, D.P., 2004. MicroRNAs modulate hematopoietic lineage differentiation. Science 303, 83–86.
Cisse, B., Caton, M.L., Lehner, M., Maeda, T., Scheu, S., Locksley, R., Holmberg, D., Zweier, C., den Hollander, N.S., Kant, S.G., Holter, W., Rauch, A., Zhuang, Y., Reizis, B., 2008. Transcription factor E2-2 is an essential and specific regulator of plasmacytoid dendritic cell development. Cell 135, 37–48.
Cobb, B.S., Hertweck, A., Smith, J., O'Connor, E., Graf, D., Cook, T., Smale, S.T., Sakaguchi, S., Livesey, F.J., Fisher, A.G., Merkenschlager, M., 2006. A role for Dicin in immune neogenesis. J. Exp. Med. 203, 2519–2527.
Corcoran, L., Ferrero, I., Vremec, D., Lucas, K., Waithman, J., O'Keeffe, M., Wu, L., Wilson, A., Shortman, K., 2003. The lymphoid past of mouse plasmacytoid cells and thymic dendritic cells. J. Immunol. 170, 4926–4932.

Edwards, A.D., Diebold, S.S., Slack, E.M., Tomizawa, H., Hemmi, H., Kaisho, T., Akira, S., Reis e Sousa, C., 2003. Toll-like receptor expression in murine DC subsets: lack of TLR7 expression by CD8 alpha+ DC correlates with unresponsiveness to imidazoquinolines. Eur. J. Immunol. 33, 827–833.
Esashi, E., Wang, Y.H., Perng, O., Qin, X.F., Liu, Y.J., Watowich, S.S., 2008. The signal transducer STAT5 inhibits plasmacytoid dendritic cell development by suppressing transcription factor IRF8. Immunity 28, 509–520.
Felli, N., Fontana, L., Pelosi, E., Botta, R., Bonci, D., Facchiano, F., Liuzzi, F., Lulli, V., Morsilli, O., Santoro, S., Valtieri, M., Calin, G.A., Liu, C.G., Sorrentino, A., Croce, C.M., Peschle, C., 2005. MicroRNAs 221 and 222 inhibit normal erythropoiesis and erythroleukemic cell growth via kit receptor down-modulation. Proc. Natl. Acad. Sci. U.S.A. 102, 18081–18086.
Filipowicz, W., Bhattacharyya, S.N., Sonenberg, N., 2008. Mechanisms of post-transcriptional regulation by microRNAs: are the answers in sight? Nat. Rev. Genet. 9, 102–114.
Fukao, T., Fukuda, Y., Kiga, K., Sharif, J., Hino, K., Enomoto, Y., Kawamura, K., Nakamura, K., Takeuchi, T., Tanabe, M., 2007. An evolutionarily conserved mechanism for microRNA-223 expression revealed by microRNA gene profiling. Cell 129, 617–631.
Gentleman, R.C., Carey, V.J., Bates, D.M., Bolstad, B., Dettling, M., Dudoit, S., Ellis, B., Gautier, L., Ge, Y., Gentry, J., Hornik, K., Hothorn, T., Huber, W., Iacus, S., Irizarry, R., Leisch, F., Li, C., Maechler, M., Rossini, A.J., Sawitzki, G., Smith, C., Smyth, G., Tierney, L., Yang, J.Y., Zhang, J., 2004. Bioconductor: open software development for computational biology and bioinformatics. Genome Biol. 5, R80.
Gilliet, M., Boonstra, A., Paturel, C., Antonenko, S., Xu, X.L., Trinchieri, G., O'Garra, A., Liu, Y.J., 2002. The development of murine plasmacytoid dendritic cell precursors is differentially regulated by FLT3-ligand and granulocyte/macrophage colony-stimulating factor. J. Exp. Med. 195, 953–958.
Griffiths-Jones, S., Saini, H.K., van Dongen, S., Enright, A.J., 2008. miRBase: tools for microRNA genomics. Nucleic Acids Res. 36, D154–D158.
Hashimi, S.T., Fulcher, J.A., Chang, M.H., Gov, L., Wang, S., Lee, B., 2009. MicroRNA profiling identifies miR-34a and miR-21 and their target genes JAG1 and WNT1 in the coordinate regulation of dendritic cell differentiation. Blood 114, 404–414.
Johnnidis, J.B., Harris, M.H., Wheeler, R.T., Stehling-Sun, S., Lam, M.H., Kirak, O., Brummelkamp, T.R., Fleming, M.D., Camargo, F.D., 2008. Regulation of progenitor cell proliferation and granulocyte function by microRNA-223. Nature 451, 1125–1129.
Landgraf, P., Rusu, M., Sheridan, R., Sewer, A., Iovino, N., Aravin, A., Pfeffer, S., Rice, A., Kamphorst, A.O., Landthaler, M., Lin, C., Socci, N.D., Hermida, L., Fulci, V., Chiaretti, S., Foa, R., Schliwka, J., Fuchs, U., Novosel, A., Muller, R.U., Schermer, B., Bissels, U., Inman, J., Phan, Q., Chien, M., Weir, D.B., Choksi, R., De Vita, G., Frezzetti, D., Trompeter, H.I., Hornung, V., Teng, G., Hartmann, G., Palkovits, M., Di Lauro, R., Wernet, P., Macino, G., Rogler, C.E., Nagle, J.W., Ju, J., Papavasiliou, F.N., Benzing, T., Lichter, P., Tam, W., Brownstein, M.J., Bosio, A., Borkhardt, A., Russo, J.J., Sander, C., Zavolan, M., Tuschl, T., 2007. A mammalian microRNA expression atlas based on small RNA library sequencing. Cell 129, 1401–1414.
Laouar, Y., Welte, T., Fu, X.Y., Flavell, R.A., 2003. STAT3 is required for Flt3L-dependent dendritic cell differentiation. Immunity 19, 903–912.
le Sage, C., Nagel, R., Egan, D.A., Schrier, M., Mesman, E., Mangiola, A., Anile, C., Maira, G., Mercatelli, N., Ciafre, S.A., Farace, M.G., Agami, R., 2007. Regulation of the p27(Kip1) tumor suppressor by miR-221 and miR-222 promotes cancer cell proliferation. EMBO J. 26, 3699–3708.
Lee, Y., Kim, M., Han, J., Yeom, K.H., Lee, S., Baek, S.H., Kim, V.N., 2004. MicroRNA genes are transcribed by RNA polymerase II. EMBO J. 23, 4051–4060.
Lewis, B.P., Burge, C.B., Bartel, D.P., 2005. Conserved seed pairing, often flanked by adenosines, indicates that thousands of human genes are microRNA targets. Cell 120, 15–20.
Lindsay, M.A., 2008. microRNAs and the immune response. Trends Immunol. 29, 343–351.
Liston, A., Lu, L.F., O'Carroll, D., Tarakhovsky, A., Rudensky, A.Y., 2008. Dicer-dependent microRNA pathway safeguards regulatory T cell function. J. Exp. Med. 205, 1993–2004.
Löffler, D., Brocke-Heidrich, K., Pfeifer, G., Stocsits, C., Hackermuller, J., Kretzschmar, A.K., Burger, R., Gramatzki, M., Blumert, C., Bauer, K., Cvijic, H., Ullmann, A.K., Stadler, P.F., Horn, F., 2007. Interleukin-6 dependent survival of multiple myeloma cells involves the Stat3-mediated induction of microRNA-21 through a highly conserved enhancer. Blood 110, 1330–1333.
Lutz, M.B., Kukutsch, N., Ogilvie, A.L., Rossner, S., Koch, F., Romani, N., Schuler, G., 1999. An advanced culture method for generating large quantities of highly pure dendritic cells from mouse bone marrow. J. Immunol. Methods 223, 77–92.
Martinez-Nunez, R.T., Louafi, F., Friedmann, P.S., Sanchez-Elsner, T., 2009. MicroRNA-155 modulates the pathogen binding ability of dendritic cells (DCs) by down-regulation of DC-specific intercellular adhesion molecule-3 grabbing non-integrin (DC-SIGN). J. Biol. Chem. 284, 16334–16342.
Mayoral, R.J., Pikin, M.E., Pachkov, M., van Nimwegen, E., Rao, A., Monticelli, S., 2009. MicroRNA-221 and 222 regulate the cell cycle in mast cells. J. Immunol. 182, 433–445.
Monticelli, S., Ansel, K.M., Xiao, C., Socci, N.D., Krichevsky, A.M., Thai, T.H., Rajewsky, N., Marks, D.S., Sander, C., Rajewsky, K., Rao, A., Kosik, K.S., 2005. MicroRNA profiling of the murine hematopoietic system. Genome Biol. 6, R71.
Muljo, S.A., Ansel, K.M., Kanellopoulou, C., Livingston, D.M., Rao, A., Rajewsky, K., 2005. Aberrant T cell differentiation in the absence of Dicer. J. Exp. Med. 202, 261–269.
Naik, S.H., Proietto, A.I., Wilson, N.S., Dakic, A., Schnorrer, P., Fuchsberger, M., Lahoud, M.H., O'Keeffe, M., Shao, Q.X., Chen, W.F., Villadangos, J.A., Shortman, K., Wu, L., 2005. Cutting edge: generation of splenic CD8+ and CD8− dendritic cell equiv-

alents in Fms-like tyrosine kinase 3 ligand bone marrow cultures. J. Immunol. 174, 6592–6597.

Naik, S.H., Sathe, P., Park, H.Y., Metcalf, D., Proietto, A.I., Dakic, A., Carotta, S., O'Keeffe, M., Bahlo, M., Papenfuss, A., Kwak, J.Y., Wu, L., Shortman, K., 2007. Development of plasmacytoid and conventional dendritic cell subtypes from single precursor cells derived in vitro and in vivo. Nat. Immunol. 8, 1217–1226.

Neilson, J.R., Zheng, G.X., Burge, C.B., Sharp, P.A., 2007. Dynamic regulation of miRNA expression in ordered stages of cellular development. Genes Dev. 21, 578–589.

Onai, N., Obata-Onai, A., Schmid, M.A., Ohteki, T., Jarrossay, D., Manz, M.G., 2007. Identification of clonogenic common Flt3+M-CSFR+ plasmacytoid and conventional dendritic cell progenitors in mouse bone marrow. Nat. Immunol. 8, 1207–1216.

Pelayo, R., Hirose, J., Huang, J., Garrett, K.P., Delogu, A., Busslinger, M., Kincade, P.W., 2005. Derivation of 2 categories of plasmacytoid dendritic cells in murine bone marrow. Blood 105, 4407–4415.

Poliseno, L., Tuccoli, A., Mariani, L., Evangelista, M., Citti, L., Woods, K., Mercatanti, A., Hammond, S., Rainaldi, G., 2006. MicroRNAs modulate the angiogenic properties of HUVECs. Blood 108, 3068–3071.

Reis e Sousa, C., 2006. Dendritic cells in a mature age. Nat. Rev. Immunol. 6, 476–483.

Robbins, S.H., Walzer, T., Dembele, D., Thibault, C., Defays, A., Bessou, G., Xu, H., Vivier, E., Sellars, M., Pierre, P., Sharp, F.R., Chan, S., Kastner, P., Dalod, M., 2008. Novel insights into the relationships between dendritic cell subsets in human and mouse revealed by genome-wide expression profiling. Genome Biol. 9, R17.

Rodriguez, A., Vigorito, E., Clare, S., Warren, M.V., Couttet, P., Soond, D.R., van Dongen, S., Grocock, R.J., Das, P.P., Miska, E.A., Vetrie, D., Okkenhaug, K., Enright, A.J., Dougan, G., Turner, M., Bradley, A., 2007. Requirement of bic/microRNA-155 for normal immune function. Science 316, 608–611.

Shigematsu, H., Reizis, B., Iwasaki, H., Mizuno, S., Hu, D., Traver, D., Leder, P., Sakaguchi, N., Akashi, K., 2004. Plasmacytoid dendritic cells activate lymphoid-specific genetic programs irrespective of their cellular origin. Immunity 21, 43–53.

Shortman, K., Liu, Y.J., 2002. Mouse and human dendritic cell subtypes. Nat. Rev. Immunol. 2, 151–161.

Shortman, K., Naik, S.H., 2007. Steady-state and inflammatory dendritic-cell development. Nat. Rev. Immunol. 7, 19–30.

Smyth, G.K., 2004. Linear models and empirical Bayes methods for assessing differential expression in microarray experiments. Stat. Appl. Genet. Mol. Biol. 3 (Article 3).

Stockinger, B., Zal, T., Zal, A., Gray, D., 1996. B cells solicit their own help from T cells. J. Exp. Med. 183, 891–899.

Suzuki, R., Shimodaira, H., 2006. Pvclust: an R package for assessing the uncertainty in hierarchical clustering. Bioinformatics 22, 1540–1542.

Taganov, K.D., Boldin, M.P., Chang, K.J., Baltimore, D., 2006. NF-kappaB-dependent induction of microRNA miR-146, an inhibitor targeted to signaling proteins of innate immune responses. Proc. Natl. Acad. Sci. U.S.A. 103, 12481–12486.

Wang, X., Seed, B., 2003. A PCR primer bank for quantitative gene expression analysis. Nucleic Acids Res. 31, e154.

Weigel, B.J., Nath, N., Taylor, P.A., Panoskaltsis-Mortari, A., Chen, W., Krieg, A.M., Brasel, K., Blazar, B.R., 2002. Comparative analysis of murine marrow-derived dendritic cells generated by Flt3L or GM-CSF/IL-4 and matured with immune stimulatory agents on the in vivo induction of antileukemia responses. Blood 100, 4169–4176.

Wu, H., Neilson, J.R., Kumar, P., Manocha, M., Shankar, P., Sharp, P.A., Manjunath, N., 2007. miRNA profiling of naive, effector and memory CD8 T cells. PLoS One 2, e1020.

Xu, Y., Zhan, Y., Lew, A.M., Naik, S.H., Kershaw, M.H., 2007. Differential development of murine dendritic cells by GM-CSF versus Flt3 ligand has implications for inflammation and trafficking. J. Immunol. 179, 7577–7584.

Zhou, X., Jeker, L.T., Fife, B.T., Zhu, S., Anderson, M.S., McManus, M.T., Bluestone, J.A., 2008. Selective miRNA disruption in T reg cells leads to uncontrolled autoimmunity. J. Exp. Med. 205, 1983–1991.

6 Publication II (Kuipers et al., J Immunol 2010)

Dicer-dependent microRNAs control maturation, function, and maintenance of Langerhans cells in vivo

Harmjan Kuipers*, Frauke M. Schnorfeil*, Hans-Jörg Fehling, Helmut Bartels and Thomas Brocker

Journal of Immunology. 2010 Jul 1;185(1):400-409[2]

* equal contribution

Article recommended by:

Miriam Merad: Faculty of 1000 Biology, 17 Jun 2010

http://f1000biology.com/article/id/3590956/evaluation

---

[2] Reprinted with permission from The Journal of Immunology. Copyright 2010. The American Association of Immunologists, Inc.

Dicer-Dependent MicroRNAs Control Maturation, Function, and Maintenance of Langerhans Cells In Vivo

Harmjan Kuipers,*,[1] Frauke M. Schnorfeil,*,[1] Hans-Jörg Fehling,[†] Helmut Bartels,[‡] and Thomas Brocker*

Dendritic cells (DCs) are central for the induction of T cell immunity and tolerance. Fundamental for DCs to control the immune system is their differentiation from precursors into various DC subsets with distinct functions and locations in lymphoid organs and tissues. In contrast to the differentiation of epidermal Langerhans cells (LCs) and their seeding into the epidermis, LC maturation, turnover, and MHC class II Ag presentation capacities are strictly dependent on the presence of Dicer, which generates mature microRNAs (miRNAs). Absence of miRNAs caused a strongly disturbed steady-state homeostasis of LCs by increasing their turnover and apoptosis rate, leading to progressive ablation of LCs with age. The failure to maintain LCs populating the epidermis was accompanied by a proapoptotic gene expression signature. Dicer-deficient LCs showed largely increased cell sizes and reduced expression levels of the C-type lectin receptor Langerin, resulting in the lack of Birbeck granules. In addition, LCs failed to properly upregulate MHC class II, CD40, and CD86 surface molecules upon stimulation, which are critical hallmarks of functional DC maturation. This resulted in inefficient induction of CD4 T cell proliferation, whereas Dicer-deficient LCs could properly stimulate CD8 T cells. Taken together, Dicer-dependent generation of miRNAs affects homeostasis and function of epidermal LCs. *The Journal of Immunology*, 2010, 185: 400–409.

Dendritic cells (DCs) are specialized for uptake, processing, and presentation of Ag and, as a consequence, do control tolerance and immunity (1). DCs develop from hematopoietic precursors into a heterogeneous population of subsets, which are distinct in their locations, phenotypes, and functions (2–4). In the absence of inflammation, two major subsets of DCs can be distinguished, conventional (cDCs) and plasmacytoid DCs (pDCs). In addition to these blood-derived cDCs and pDCs, tissue-derived migratory DCs can be found in lymph nodes. In skin-draining lymph nodes (sLNs), this group of migratory DCs mainly consists of Langerin-expressing epidermal Langerhans cells (LCs) as well as Langerin+ and Langerin− dermal DCs cells that migrated from the skin (5–7).

The complexity of different DC subtypes and their relative functions was elucidated only recently. Their central role in the maintenance of immune homeostasis was demonstrated, for example, by constitutive ablation of DCs in vivo, which resulted

*Institute for Immunology and †Institute for Anatomy, Ludwig-Maximilian-University Munich, Munich; and ‡Institute of Immunology, University Clinics Ulm, Ulm, Germany

[1]H.K. and F.S. contributed equally to this work.

Received for publication December 7, 2009. Accepted for publication April 26, 2010.

This work was supported by Deutsche Forschungsgemeinschaft Grant KU 2513/1-1 (to H.K. and T.B.) as well as Rubicon Grant 825.06.013 from the Dutch Scientific Organization (to H.K.).

The microarray data presented in this article have been submitted to the ArrayExpress database under accession number E-MEXP-2587.

Address correspondence and reprint requests to: Prof. Thomas Brocker, Institute for Immunology, Ludwig-Maximilian-University Munich, Goethestrasse 31, 80336 Munich, Germany. E-mail address: brocker@lmu.de

The online version of this article contains supplemental material.

Abbreviations used in this paper: cDC, conventional dendritic cell; CHS, contact hypersensitivity; DC, dendritic cell; DNFB, 2,4-dinitro-1-fluorobenzene; EpCAM, epithelial cell adhesion molecule; gB, glycoprotein B; GO, Gene Ontology; LC, Langerhans cell; miRNA, microRNA; N.D., not detectable; pDC, plasmacytoid dendritic cell; qPCR, quantitative PCR; RFP, red fluorescent protein; sLN, skin-draining lymph node.

Copyright © 2010 by The American Association of Immunologists, Inc. 0022-1767/10/$16.00

www.jimmunol.org/cgi/doi/10.4049/jimmunol.0903912

in spontaneous fatal autoimmunity (8). In addition, induction of T cell immunity also depends on DCs (9). A division of labor can be observed for certain tissue-derived DCs, such as LCs, which may induce potent T cell immunity either directly (10) or indirectly by transporting Ag from skin to lymph nodes for transfer to CD8a+ resident DCs. The latter then cross-present the transferred Ag from skin-derived viral infections to CD8 T cells (11). In contrast, CD8− DCs preferentially prime CD4 T cells via MHC class II presentation (12). Moreover, DCs mediate peripheral T cell tolerance by deleting self-reactive CD8 T cells, which otherwise could induce autoimmune reactions (13). Despite this striking complexity of different phenotypes and functions, DCs seem to differentiate from common precursors by largely unknown mechanisms. Cytokines, particularly flt3, and transcription factors have been shown to contribute to this process (4, 14, 15).

A novel control mechanism of immune homeostasis has become evident with the discovery of microRNAs (miRNAs), which are noncoding RNA molecules of an average of 21–22 nt in length (16). miRNAs bind to the 3′ untranslated region of mRNAs, resulting in inhibition of translation or degradation of mRNA, thereby providing mechanisms for posttranscriptional regulation of gene expression. A key enzyme in the miRNA biogenesis pathway is the endonuclease Dicer, which generates 22-nt mature miRNAs. Therefore, the deficiency for the gene encoding Dicer, *dicer1*, abolishes generation of miRNAs and their regulatory functions.

It has been shown that a large number of specific miRNAs are involved in immunological processes, with major roles for miR-150 and miR-155 in lymphocyte differentiation and function (17, 18). Furthermore, several miRNAs have been identified that are involved in human monocyte differentiation to DCs in vitro (19–21). Dicer has also been implicated in several immunological cell lineages, such as T cells (22–24), regulatory T cells (25–27), or B cells (28), with drastic consequences for development and functions of the relevant cells. In contrast, nothing is known about the role of Dicer in DCs in vivo. In this report, we analyzed development and function of Dicer-deficient DCs. Upon conditional deletion of *dicer1* in CD11c$^+$ DCs, we could not detect an effect of Dicer deficiency on short-lived resident DC subtypes in spleen and lymph nodes. However, the dense LC network in the epidermis could not be maintained in the absence of Dicer. In addition, LCs displayed a defect in maturation and Ag presentation. These data identify Dicer-generated miRNAs as key regulators of LC homeostasis and function.

Materials and Methods
Mice

To generate mice deficient in mature miRNAs in DCs, previously described Dicer$^{fl x/fl x}$ mice (29) on a mixed C57BL/6 background were crossed with CD11c-Cre mice (30) (CD11c-Cre-Dicer$^{fl/fl}$). For tracking Cre expression, CD11c-Cre-Dicer$^{fl/fl}$ mice were bred to ROSA26-tdRFP mice (CD11c-Cre-tdRFP-Dicer$^{fl/fl}$) (31). OT-I and OT-II mice (expressing a transgenic TCR specific for OVA$_{257-264}$/MHC H-2K^b or OVA$_{323-339}$/MHC H-2A^b, respectively) were originally obtained from The Jackson Laboratory (Bar Harbor, ME). Mice were bred and maintained in a conventional facility at the Institute of Immunology (Munich, Germany) and used according to protocols approved by the local animal ethics committee.

Gene expression analysis

To analyze recombination of *dicer1* alleles in sorted CD11c$^+$ splenocytes, genomic DNA was extracted using a commercial kit according to the manufacturer's instructions (Qiagen, Hilden, Germany) and genotyping PCR was performed as described (29). To quantify mRNA expression levels, RNA was isolated from sorted red fluorescent protein (RFP)$^+$ spleen DCs using a RNeasy Mini Kit (Qiagen) or from isolated LCs using a RNAqueous-Micro Kit (Ambion, Darmstadt, Germany). cDNA was generated with SuperScript III First-Strand Synthesis System using random primers (Invitrogen, Darmstadt, Germany). Quantitative PCR reactions were run in a CFX96 PCR machine (Bio-Rad, Munich, Germany) using the LightCycler TaqMan Master Kit (Roche, Mannheim, Germany) and gene-specific primers (Applied Biosystems, Darmstadt, Germany) according to the manufacturer's instructions. Expression levels were normalized to ubiquitin c, and relative expression was calculated using the $\Delta\Delta C_T$ method. For quantitative real-time RT-PCR of miRNAs, total RNA including the low m.w. fraction was isolated from RFP$^+$ spleen DCs or LCs using a miRNeasy Kit (Qiagen) or a RNAqueous-Micro Kit (Ambion) modified for recovery of small RNAs according to manufacturer's instructions. Gene-specific reverse transcription was performed for each miRNA using the TaqMan MicroRNA Reverse Transcription Kit and gene-specific primers (Applied Biosystems) following the manufacturer's protocol. Quantitative PCR (qPCR) reactions were run in a CFX96 PCR machine (Bio-Rad) using a LightCycler TaqMan Master Kit (Roche) according to the manufacturer's instructions. Expression levels were normalized to RNU19, and relative expression was calculated using the $\Delta\Delta C_T$ method.

Flow cytometry and cell purification

The mAbs used were FITC-, PE-, allophycocyanin-, PECy5.5-, Alexa Fluor 647-, or PerCP-conjugated anti-mouse I-Ab, H-2K^b, CD49b, F4/80, CD11c, PDCA-1, Langerin, CD24, CD40, CD45, CD86, CCR7, and epithelial cell adhesion molecule (EpCAM) (eBioscience, Frankfurt, Germany and BD Biosciences, Heidelberg, Germany). H-2K^b/HSV-glycoprotein B (gB)$_{498-505}$ tetramers were purchased from ProImmune (Oxford, U.K.). Single-cell suspensions of lymphoid organs were obtained by Liberase CI (0.42 mg/ml; Roche) and DNaseI (0.2 mg/ml; Roche) treatment. Fc block (anti-FcγRII/III Ab) was included in every staining. To separate epidermal and dermal sheets, mouse ears were split into dorsal and ventral parts and incubated for 60 min at 37°C in PBS containing 0.5% trypsin and 5 mM EDTA (5). Epidermal sheets were subsequently incubated for 2.5 h in collagenase 4 (1.5 mg/ml; Worthington, Lakewood, NJ), whereas dermal sheets were further processed using an enzyme mixture as described by Bursch et al. (6). Intracellular staining against Langerin was performed with the BD Cytofix/Cytoperm reagents (BD Biosciences) according to the manufacturer's protocol. To analyze LC maturation, epidermal sheets were incubated for 48 h at 37°C on complete RPMI media and emigrated cells were analyzed by flow cytometry. To obtain OT-I and OT-II T cells, cell suspensions of spleens and lymph nodes of TCR transgenic mice were prepared. OT-I and OT-II T cells were purified using a CD8$^+$ or CD4$^+$ T cell Isolation Kit (Miltenyi Biotec, Bergisch Gladbach, Germany), respectively, following the manufacturer's protocol. For all of the experiments, data were acquired on FACSCanto2 or sorted on FACSAria flow cytometers (BD

Biosciences) to 88–98% purity after gating out dead cells. Flow cytometric data were analyzed using FlowJo software (Tree Star, Ashland, OR).

Western blotting

Approximately 0.5×10^6 CD11c$^+$ sorted splenocytes were lysed in buffer containing 1% Nonidet P-40 and protease inhibitors and loaded onto a 4–15% gradient gel (Bio-Rad). To detect Dicer protein on Western blots, Dicer 1416 (a kind gift from the D. Livingston laboratory, Dana Farber Cancer Institute, Boston, MA) was used. To control for cell loading, blots were subsequently stained with anti-β-actin Ab (Sigma-Aldrich, Munich, Germany). Bound primary Abs were detected with appropriate peroxidase-conjugated secondary Abs (Dako, Hamburg, Germany and Amersham Biosciences, Freiburg, Germany), and signal intensities were quantified with ImageJ software (National Institutes of Health, Bethesda, MD).

In vivo cytotoxic T cell assay

Erythrocyte-depleted C57BL/6 splenocytes were pulsed with 2 μg/ml HSV-gB$_{498-505}$ peptide (SSIEFARL) for 2 h at 37°C and subsequently labeled with 1.67 μM CFSE (CFSE^{hi} population). Unpulsed control cells were labeled with 0.07 μM CFSE (CFSE^{low} population). A total of 5×10^6 cells of each CFSE-labeled cell cohort were mixed and injected i.v. Twenty-four hours later, mice were sacrificed and spleens were analyzed by flow cytometry.

ELISA

For the detection of HSV-specific Abs in the serum, ELISA was performed as described (32), using peroxidase-labeled second-step Abs specific for mouse total IgG or IgG2a.

Immunofluorescence

Mouse ears were mechanically split into dorsal and ventral halves and incubated in 0.5 M ammonium thiocyanate (Sigma-Aldrich) for 20 min at 37°C to allow for separation of epidermal sheets from the dermis. After fixation in acetone for 5 min at room temperature, sheets were blocked in PBS containing 0.25% BSA and 10% mouse serum and stained with biotin-conjugated anti–I-A/I-E, followed by Alexa Fluor 555-conjugated streptavidin (Invitrogen). Sheets were then mounted in Fluoromount (Southern Biotechnology Associates, Birmingham, AL) and analyzed on an BX41TF-5 microscope equipped with a F-View II Digital camera and CELL-BND-F software (Olympus, Hamburg, Germany).

Electron microscopy

Ear tips were fixed in a solution of 3.5% glutaraldehyde or 2.5% glutaraldehyde and 2% formaldehyde, freshly prepared from paraformaldehyde, in 0.1 M sodium cacodylate-HCl buffer (pH 7.3). The tissue specimens were cut in small blocks of 1–2 mm in length and, after thorough washing in the buffer, postfixed in 2% OsO$_4$ in 0.1 M sodium cacodylate-HCl buffer, dehydrated in ethanol, and embedded in Araldite (Ted Pella, Redding, CA). Thin sections were stained with lead citrate and uranyl acetate and examined in a CM10 electron microscope (Philips, Eindhoven, The Netherlands) at 80 kV. To determine the occurrence of Birbeck granules, LCs were investigated in three blocks per animal of the two groups. A total of 100 LCs in the Dicer$^{wt/wt}$ mice and of 60 LCs in Dicer$^{Δ/Δ}$ mice were examined.

Contact hypersensitivity response

To induce ear swelling, mice were sensitized with either 25 μl 0.3% 2,4-dinitro-1-fluorobenzene (DNFB) or 50 μl 0.5% Oxazolone (both Sigma-Aldrich) in acetone/olive oil (4:1) on the shaved abdomen and on day 5 challenged with 5 μl 0.15% DNFB or 10 μl 0.25% Oxazolone in acetone/olive oil (4:1), respectively, on both sides of one ear. Swelling was measured by comparing ear thickness before and 24 h after challenge using a micrometer (Mitutoyo, Eisenach, Germany).

Migration assay

Mouse ears were mechanically split into ventral and dorsal halves, and the dorsal halves were floated dermal side down on complete RPMI media containing 0.1 μg/ml CCL21 (R&D Systems, Wiesbaden-Nordenstadt, Germany). Ear halves were cultured for 3 d at 37°C, and after 24, 48, and 72 h, migratory DCs were collected and halves were transferred to fresh culture media. Collected cells were pooled and stored at 4°C until analysis on day 3 via flow cytometry.

Ag presentation assay

Epidermal sheets were prepared as described and incubated with 0.25 mg/ml OVA protein (Sigma-Aldrich) overnight. Migrated cells and sheets were washed and incubated further in fresh medium. At day 2 of culture, migrated cells were collected, and light-density LCs were enriched using OptiPrep (Axis-Shield, Oslo, Norway) in a density gradient centrifugation. Equal numbers of LCs were seeded into the wells of a round-bottom 96-well plate and cocultured with 2.5 × 10⁴ CFSE-labeled OT-I or OT-II T cells in complete RPMI medium containing GM-CSF. T cell proliferation was analyzed by flow cytometry after 60–65 h of culture.

BrdU and TUNEL labeling

Mice were injected with 1 mg BrdU i.p. and kept on 0.8 mg/ml BrdU-supplemented water for 7 d. Epidermal cell suspensions were stained for surface markers and BrdU incorporation using the BrdU Flow Kit (BD Biosciences) following the manufacturer's protocol. For apoptosis detection by TUNEL assay (In Situ Cell Death Detection Kit; Roche), epidermal cell suspensions were stained for surface markers. Cells were fixed, permeabilized, and enzymatically labeled with fluorescein–dUTP according to the manufacturer's instructions. The percentage of LCs that incorporated BrdU or dUTP, respectively, was determined by flow cytometry.

Statistical analysis

Data were analyzed using the unpaired, two-tailed Student t test. A p value <0.05 was considered to be significant. For analysis of fluorescence intensities, the median fluorescence intensity was calculated.

Results

Characterization of Dicer-deficient DCs

To study the role of Dicer and miRNAs in DCs, we crossed mice containing loxP-flanked $dicer1$ alleles (floxed; Dicer$^{fl/fl}$) (29) with CD11c-Cre transgenic mice expressing Cre in CD11c⁺ DCs (30). The resulting CD11c-Cre-Dicer$^{fl/fl}$ (Dicer$^{Δ/Δ}$) mice did not exhibit any obvious abnormalities, were viable and fertile, and were born at the expected Mendelian ratio (data not shown). We first determined whether the $dicer1$ alleles were effectively excised from the DC genome and analyzed expression of Dicer transcripts and protein from sorted CD11c⁺ spleen DCs. PCR on genomic DNA from FACS-sorted CD11c⁺ splenocytes showed that the loxP-flanked alleles (fl/fl) were efficiently recombined in Cre-positive cells (Δ/Δ) (Fig. 1A). This deletion led to disappearance of Dicer mRNA (Fig. 1B). qPCR analysis of the deletion of Dicer mRNA indicated 99.9% efficiency of deletion. Consequently, also Dicer protein was present only at minimal amounts, as detected by Western blot (Fig. 1C). However, qPCR quantification of some mature miRNAs, which are known to be expressed in DCs (33; data not shown), revealed only modestly reduced miRNA levels in Dicer-deficient

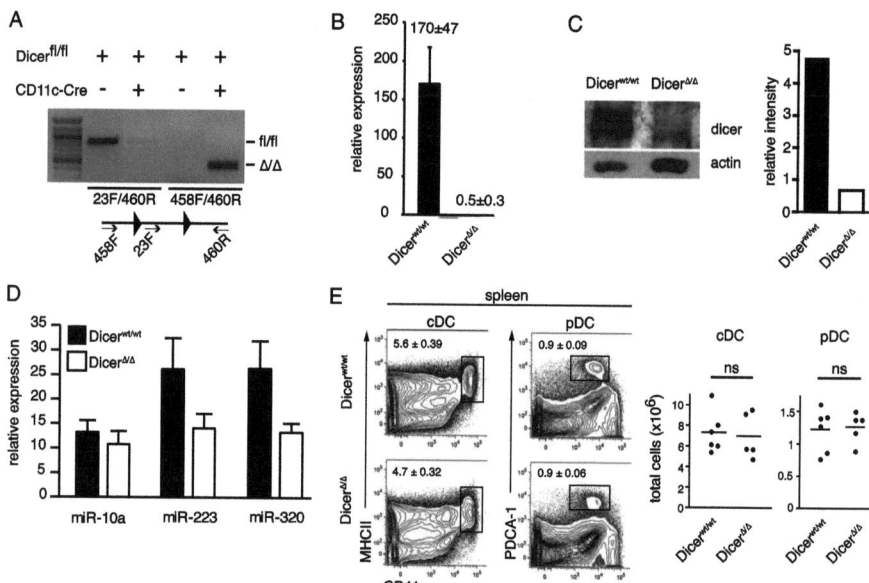

FIGURE 1. Dicer deletion and reduction of miRNA levels do not affect splenic DC populations in CD11c-Cre-Dicer$^{fl/fl}$ mice. *A*, Genotyping PCR on FACS-sorted CD11c⁺ splenocytes isolated from CD11c-Cre-Dicer$^{fl/fl}$ and Dicer$^{fl/fl}$ mice. Presence of floxed $dicer$ alleles and Cre recombinase in the sorted cells is indicated above the gel image. Primer pairs and amplification strategy to discriminate floxed alleles from recombined alleles (Δ/Δ) are depicted below the gel image. Bold arrows denote loxP recombination sites; small arrows denote PCR primers. *B*, qPCR quantification of Dicer mRNA levels in FACS-sorted RFP⁺ spleen DCs from control (Dicer$^{wt/wt}$) and conditionally deleted Dicer (Dicer$^{Δ/Δ}$) animals that were bred to ROSA26-tdRFP reporter mice. Expression levels were normalized to ubiquitin c reference gene levels. Error bars denote SEM ($n = 4$). *C*, Detection of Dicer protein by Western blotting of cell lysates from FACS-sorted CD11c⁺ splenocytes from control (Dicer$^{wt/wt}$) and conditional deleted Dicer (Dicer$^{Δ/Δ}$) animals. As a loading control, actin was detected. Relative signal intensities were quantified by normalizing Dicer to actin staining intensities. *D*, Analysis of miRNA levels in RFP⁺ spleen DCs from Dicer$^{wt/wt}$ and Dicer$^{Δ/Δ}$ animals by qPCR. miRNA expression levels were normalized to RNU19 reference gene levels. Error bars denote SEM ($n = 4$). *E*, FACS analysis of DC populations in spleens obtained from Dicer$^{wt/wt}$ and Dicer$^{Δ/Δ}$ mice. Percentages of cells that fall into each gate (mean ± SEM, $n = 5–6$) are indicated. All of the populations depicted were gated on live cells; NK cells and macrophages were gated out by staining with CD49b and F4/80, respectively. Data were combined from two independent experiments with similar results.

DCs (Fig. 1D). For example, miR-223 and miR-320 were ~2-fold reduced in DCs from Dicer$^{\Delta/\Delta}$ mice, whereas we could not detect reduced expression of miR-10a. Consequently, the frequencies and total cell counts of splenic cDCs (CD11c^{high}MHC II$^+$) and pDCs (CD11c^{int}PCDA-1$^+$) were not significantly different between both mouse strains (Fig. 1E), and these cells showed normal surface expression of typical surface markers (data not shown). Further characterization of cDC subsets showed that also the frequencies of CD8α^+ and CD8α^- DCs were not significantly altered (data not shown). The numbers of other leukocyte cell populations from spleens, such as T and B cells, NK cells, macrophages, thymocytes, and immature as well as mature neutrophils, were also similar between Dicer$^{\Delta/\Delta}$ and CD11c-Cre control (Dicer$^{wt/wt}$) mice (Supplemental Fig. 1). To test if this apparent normality of DC phenotype and differentiation was eventually accompanied by functional defects, we infected mice i.v. with HSV. However, Dicer$^{\Delta/\Delta}$ mice mounted normal HSV-gB–specific cytotoxic T cell responses as detected with the respective MHC multimers (Fig. 2A). These CD8$^+$ HSV-gB–specific CTLs were functionally indistinguishable from those found in wild-type mice as revealed

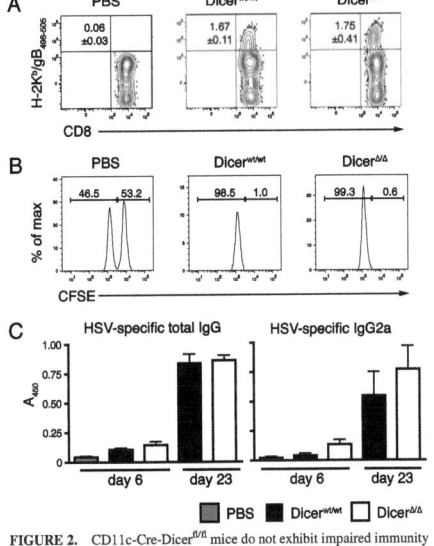

FIGURE 2. CD11c-Cre-Dicer$^{fl/fl}$ mice do not exhibit impaired immunity to systemic viral infection. At day 0, Dicer$^{wt/wt}$ ($n = 5$) and Dicer$^{\Delta/\Delta}$ ($n = 5$) mice were infected systemically with HSV-1 strain KOS (i.v. 1 × 10^5 PFU). Control mice ($n = 5$) received PBS i.v. A, The frequency of HSV-gB–specific CD8 T cells in peripheral blood at day 10 postinfection was determined by staining with gB$_{498-505}$–MHC tetramer complexes. Peripheral blood cells were acquired by flow cytometry and gated on CD8$^+$ lymphocytes. FACS plots are representative for a sample from each group. Numbers denote frequency of H-2K^b/gB$_{498-505}$–positive cells as percentage of CD8$^+$ T cells ± SEM. B, On day 25 postinfection, splenocytes were subjected to an in vivo cytotoxic T cell assay. A mixture of HSV-gB–SSIEFARL peptide-loaded target cells (CFSE^{high}) and control cells (CFSE^{low}) was injected i.v., and the specific lysis was analyzed by flow cytometry 24 h later. A representative plot from each group is shown. Numbers denote frequency of cells in each gate as a percentage of the total CFSE-labeled population. C, On days 6 and 23, sera were analyzed for HSV-specific total IgG and IgG2a Abs using ELISA. Error bars denote SEM.

by in vivo cytotoxic assays (Fig. 2B). In addition, Dicer$^{\Delta/\Delta}$ mice generated identical HSV-specific Ab responses as compared with wild-type control animals (Fig. 2C). Because DCs are necessary for the induction of antiviral T cell and B cell responses, these findings suggest that DCs from Dicer$^{\Delta/\Delta}$ mice are not functionally defective.

CD11c-mediated Dicer ablation results in reduction of specific DC subsets

To further assess the phenotype of DCs in the absence of Dicer, we performed flow cytometric analysis of DC subsets in lymph nodes. We could not detect differences in cDCs of mesenteric lymph nodes (Fig. 3A). The cDC population in sLNs can be further subdivided into blood-derived, resident DCs (CD11c^{high}MHC II$^+$) and migratory DCs (CD11c^{int}MHC II^{high}) consisting of epidermal LCs and dermal DCs immigrating from skin. This migratory DC subgroup from skin was significantly decreased in Dicer$^{\Delta/\Delta}$ mice (Fig. 3A). To differentiate the various DC populations from skin, we performed further characterization by flow cytometry. Using mAbs specific for MHC class II, Langerin, and EpCAM, we differentiated between LCs and Langerin$^+$ or Langerin$^-$ dermal DCs. As shown in Fig. 3B, we found a 12-fold reduction of MHC II$^+$Langerin$^+$ LCs in the epidermis of CD11c-specific Dicer$^{\Delta/\Delta}$ animals ($p = 0.0032$). Consequently, also the transmigrating LCs found in underlying dermal regions were nearly undetectable (Fig. 3B, dermis). In contrast, the other two main dermal DC subpopulations, the Langerin$^+$ and Langerin$^-$ dermal DCs, which are both EpCAM$^-$, were present at normal frequencies (Fig. 3B).

To ensure that the few LCs found in the epidermis of Dicer$^{\Delta/\Delta}$ mice were actually expressing Cre and would not be cells escaping the loxP rearrangement due to lack of Cre expression, we analyzed the LCs from epidermal sheets of CD11c-Cre-tdRFP-Dicer$^{fl/fl}$ mice. Despite significantly lower numbers of LCs in the epidermis of CD11c-Cre-tdRFP-Dicer$^{fl/fl}$ mice as compared with those found in CD11c-Cre-tdRFP-Dicer$^{wt/wt}$ littermates, the percentages of RFP$^+$ LCs were nearly identical (Fig. 3C) and therefore indicative for Cre expression. As a consequence, only background levels of Dicer mRNA could be detected by qPCR in sorted LCs from Dicer$^{\Delta/\Delta}$ mice (Fig. 3D). This indicated that the few LCs found in Dicer$^{\Delta/\Delta}$ mice actually deleted *dicer1* with a deletion efficacy of 99.1%. Also, the analysis of miRNA-223 indicated a deletion efficacy of 93.2% for this miRNA in LCs (Fig. 3D). The Dicer-deficient remaining LCs showed ~30% enlarged cell sizes as indicated by an increased forward scatter signal during flow cytometry (Fig. 3E; $p = 0.0002$). Steady-state surface expression levels of other typical LC markers, such as MHC class II and CD24 (Fig. 3C), and CD11c, CD40, CD86, and CCR7 (see below, Fig. 6B) were unaltered as compared with those of Dicer$^{wt/wt}$ mice. However, Dicer-deficient LCs expressed only half the amount of Langerin as compared with that of their wild-type counterparts (Fig. 3E; $p = 0.0057$) as well as substantially decreased levels of TGF-βRII (Fig. 3E).

Dicer-deficient Langerhans cells do not develop Birbeck granules and are lost with age

This selective loss of LCs from the skin was also confirmed by histological analysis of epidermal sheets (Fig. 4A). In 8- to 9-wk-old mice, Dicer$^{\Delta/\Delta}$ LCs were distributed sparsely but equally in the epidermis. Quantification of LCs revealed ~12-fold reduction of Dicer$^{\Delta/\Delta}$ LCs (Fig. 4C), similarly to our observations by flow cytometry (Fig. 3B). The morphology of the remaining Dicer$^{\Delta/\Delta}$ LCs was different, because they occupied a larger area and had longer dendrites (Fig. 4A). It remains to be seen whether this change of cell shape is directly correlated to the loss of *dicer1* expression or

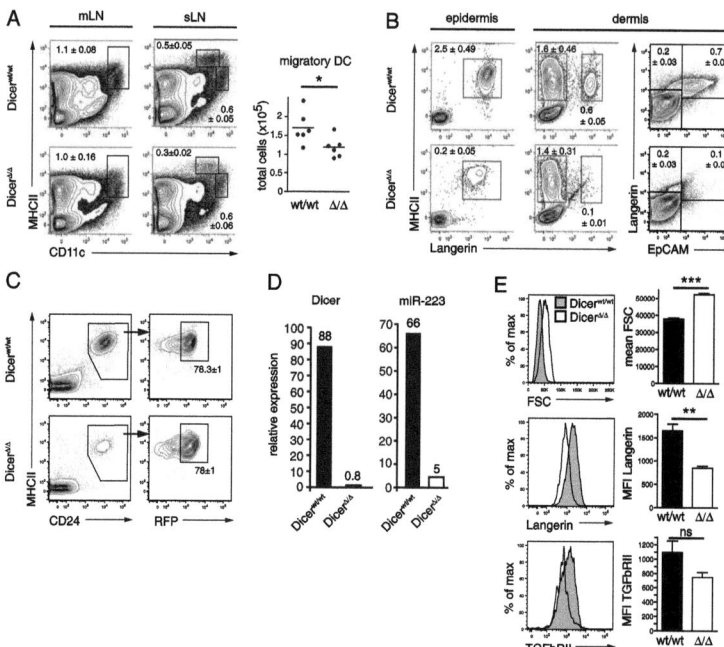

FIGURE 3. CD11c-mediated deletion of Dicer results in the absence of specific DC populations in peripheral lymphoid organs and in the skin. *A*, FACS analysis of DC populations in mesenteric lymph nodes and sLNs obtained from control Dicer$^{wt/wt}$ and Dicer$^{\Delta/\Delta}$ mice. Percentages of cells that fall into each gate are indicated (mean ± SEM, n = 5–6). All of the populations depicted were gated on live cells; NK cells and macrophages were gated out by staining with CD49b and F4/80, respectively. Combined data from two independent experiments with similar results are shown (*p < 0.05). *B*, FACS analysis of single-cell suspensions from epidermis and dermis obtained from Dicer$^{wt/wt}$ and Dicer$^{\Delta/\Delta}$ animals. Epidermal and dermal cell populations shown are gated on CD45$^+$ cells or, for analysis of dermal cell populations based on Langerin and EpCAM, on CD45$^+$MHC II$^+$ cells. Numbers on plots indicate cells in each gate as a percentage of CD45$^-$ cells (mean ± SEM; n = 4). *C*, FACS analysis of CD45$^+$-gated epidermal cells derived from Dicer$^{wt/wt}$ and Dicer$^{\Delta/\Delta}$ animals. Percentages of RFP$^+$ LCs are indicated (mean ± SEM; n = 3). *D*, qPCR analysis of Dicer mRNA and miR-223 miRNA levels in FACS-sorted LCs pooled from 3 Dicer$^{wt/wt}$ or 14 Dicer$^{\Delta/\Delta}$ animals. Values indicated represent relative expression levels in LCs from Dicer$^{\Delta/\Delta}$ and Dicer$^{wt/wt}$ animals. Expression was normalized to ubiquitin c (Dicer) or RNU19 (miR-223). *E*, CD45$^+$MHC II$^+$Langerin$^+$ LCs from Dicer$^{wt/wt}$ and Dicer$^{\Delta/\Delta}$ animals were assessed for their size by forward scatter measurement (*top panel*) and their Langerin (*middle panel*) and TGFβRII (*bottom panel*) expression levels by FACS staining. Representative histogram overlays of Dicer$^{wt/wt}$ (gray) and Dicer$^{\Delta/\Delta}$ (open) LC measurements are shown. Graphs indicate mean ± SEM (n = 3). Data are representative of at least three independent experiments. **p < 0.01; ***p < 0.001.

a consequence of decreased LC density. In contrast, no differences in other leukocyte populations of the skin, such as dendritic epidermal T cells (CD45$^+$TCRγδ$^+$), could be detected (data not shown). The loss of LCs was increasing with age, because young Dicer$^{\Delta/\Delta}$ mice at the age of 10 d still contained normal numbers of LCs (Fig. 4*B*, 4*C*). This was not due to lack of Cre expression in the animals, because 10-d-old CD11c-Cre-tdRFP-Dicer$^{fl/fl}$ mice had already uniformly RFP$^+$ LCs in the skin (data not shown). In contrast, with increasing age, their LC numbers decreased drastically, and at 6 mo of age hardly any LCs could be found (Fig. 4*B*, Dicer$^{\Delta/\Delta}$, 6 mo). The extremely few LCs in old mice were mostly clustered in "patches" (Fig. 4*B*, Dicer$^{\Delta/\Delta}$, 6 mo). In the epidermis of 18-mo-old animals, only extremely few LCs could be found, among them some LCs with giant extremities (Fig. 4*B*, *bottom panel*).

To further analyze LCs, we performed electron microscopy studies. LCs were distinguished from keratinocytes in thin sections by the presence of Birbeck granules and the absence of thick bundles of intermediate keratin filaments and desmosomes (Fig. 4*D*). Although the frequency of LCs was not studied quantitatively with this method, these cells were clearly less frequently present in the epidermis of Dicer$^{\Delta/\Delta}$ mice as compared with control Dicer$^{wt/wt}$ animals, confirming our analysis by flow cytometry (Fig. 3*B*) and histology (Fig. 4*A*, 4*B*). Whereas Birbeck granules were present in 87% of randomly sectioned LCs in the Dicer$^{wt/wt}$ mice, they were observed in only ∼7% of the LCs examined in the Dicer$^{\Delta/\Delta}$ mice. Taken together, ablation of Dicer led to a progressive loss of LCs in the epidermis with time, with some remaining cells that showed altered morphology, lower expression levels of Langerin (Fig. 3*E*), and lack of Birbeck granules (Fig. 4*D*).

Dicer-deficient Langerhans cells have a proapoptotic gene expression signature

Given the pivotal function of Dicer in miRNA biogenesis, we expected that absence of Dicer in LCs would result in profound

The Journal of Immunology

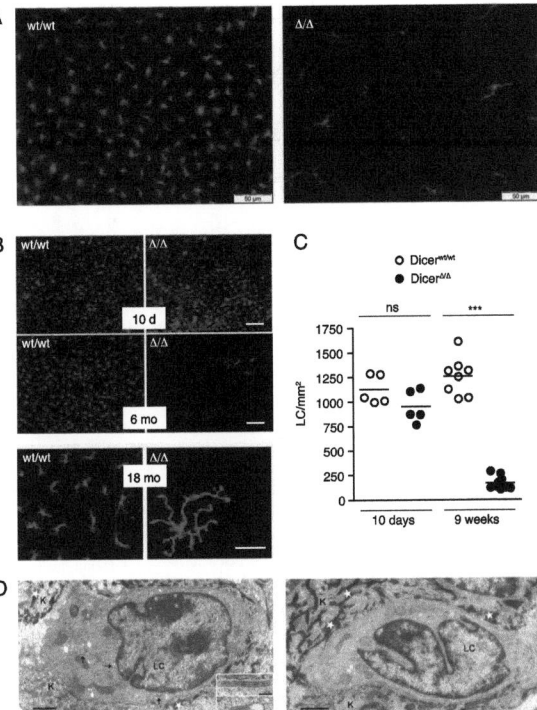

FIGURE 4. Progressive loss and altered morphology of LCs in the absence of Dicer. *A* and *B*, Immunofluorescence analysis using Alexa Fluor 555-labelled MHC II–mAb of epidermal sheets from (*A*, original magnification ×40) 9-wk-old or (*B*, original magnification ×20) 10-d-, 6-, or 18-mo-old Dicer$^{wt/wt}$ and Dicer$^{\Delta/\Delta}$ animals are shown. The skin of 6-mo-old and older mice barely contained LCs. The rare areas containing LCs were chosen for display (10 d and 6 mo, scale bar, 100 μm; 18 mo, scale bar, 50 μm). *C*, Quantification of LCs in epidermal sheets at the indicated age. The graph shows the mean of at least five fields counted from at least two mice per genotype (***$p < 0.0001$). *D*, Electron microscopy images of the epidermis from Dicer$^{wt/wt}$ and Dicer$^{\Delta/\Delta}$ animals show LCs surrounded by keratinocytes with bundles of keratin filaments (asterisks). The arrows point at three Birbeck granules, two of which are enlarged in the *insets* (original magnification ×6500; *insets*, ×39000). Note the absence of Birbeck granules in the LCs of the Dicer$^{\Delta/\Delta}$ mouse. Scale bars, 1 μm; *inset* scale bar, 0.1 μm.

changes in gene expression profiles. To determine whether individual genes or gene sets were differentially expressed between Dicer-deficient and wild-type LCs, we purified LCs from epidermal layers of the skin and analyzed gene expression using microarrays. Due to the extremely low numbers of LCs in Dicer-deficient animals, we had to include amplification steps before analysis. However, we could not detect individual genes that were differentially expressed with statistical significance (data not shown). Gene set enrichment analysis using Gene Ontology (GO) biological process terms identifies overrepresentation of genes belonging to a given functional GO term in an ordered list of genes. Differences in biological processes could be revealed this way, and we therefore analyzed the differentially expressed gene list for enrichment of genes belonging to biological process GO terms. The amount of overrepresentation is assessed with a statistical score. Two enriched GO terms, "induction of apoptosis by intracellular signals" (GO.ID: 0008629; $p = 0.0145$) and "regulation of caspase activity" (GO.ID: 0043281; $p = 0.0111$) contain proapoptotic genes. Also the proapoptotic gene *bcl2l11* (encoding for Bim), belonging to the enriched GO category "postembryonic organ development" (GO.ID: 0048569; $p = 0.0064$), is differentially regulated in Dicer$^{\Delta/\Delta}$ LCs. Verification of differential gene expression by qPCR on mRNA samples from LCs proved to be very difficult due to the extremely low numbers of Dicer$^{\Delta/\Delta}$ LCs in the skin. Nevertheless, pooling the few Dicer$^{\Delta/\Delta}$ LCs from the skin of dozens of mice allowed the isolation of sufficient amounts of mRNA for the verification of a few genes (Fig. 5*A*). This analysis showed that the expression of *dicer1* in the purified Dicer$^{\Delta/\Delta}$ LCs was reduced to >99% as compared with Dicer-sufficient LCs and served as a quality control (Fig. 5*A*). Furthermore, genes predicted by the GO term analysis, such as *myc* and *bcl2l11* (*bim*), were indeed expressed differentially in Dicer$^{\Delta/\Delta}$ LCs (Fig. 5*A*). The absence of Dicer reduced the expression of Myc transcripts ~17-fold in LCs, whereas the expression levels of Bim increased 5.3-fold (Fig. 5*A*). Altogether, these data suggest that Dicer-deficient LCs might be engaged in active apoptosis.

To test this hypothesis, we performed in vivo BrdU labeling and TUNEL assays to determine turnover and cell death of LCs. During the 7-d period of BrdU labeling, ~13% of wild-type LCs did incorporate BrdU (Fig. 5*B*). This rate of BrdU incorporation indicates the total turnover time for wild-type LCs to be ~53.8 d and confirms previous publications that estimated 53–78 d as turnover time of LCs in absence of inflammation (reviewed in Ref. 34). In marked contrast, 36% of Dicer-deficient LCs incorporated BrdU (Fig. 5*B*), indicating a nearly three times faster turnover of only 19.4 d. TUNEL assay revealed an ~7-fold increased rate of apoptosis in Dicer$^{\Delta/\Delta}$ LCs as compared with that in Dicer-sufficient LCs (Fig. 5*B*). Therefore, ablation of Dicer leads to modulation of gene expression involved in apoptosis of LCs. This enhanced the turnover and increased the apoptosis rate, resulting in progressive loss of Dicer$^{\Delta/\Delta}$ LCs. From these results, we conclude that miRNAs

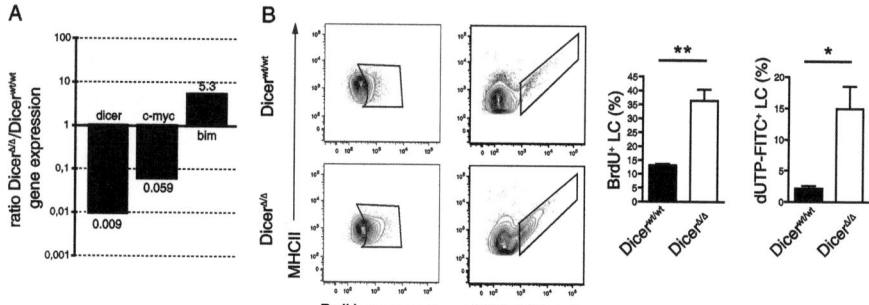

FIGURE 5. Dicer-deficient LCs display an apoptotic expression signature and increased turnover. *A*, qPCR analysis of genes predicted to be differentially regulated by microarray analysis. FACS-sorted LCs were pooled from 3 Dicer$^{wt/wt}$ or 14 Dicer$^{\Delta/\Delta}$ animals. Values indicate ratios of Dicer$^{\Delta/\Delta}$/Dicer$^{wt/wt}$ gene expression levels. Expression was normalized to ubiquitin c reference gene levels. *B*, CD45^{+}MHC II^{+} LCs in single-cell suspensions from epidermis of 7 d BrdU-fed mice were analyzed by flow cytometry for BrdU incorporation (*left panels*). TUNEL-staining of LCs is shown (*right panels*). Graphs indicate the mean ± SEM (*n* = 3–4). ∗*p* < 0.05; ∗∗*p* < 0.005.

regulate key cellular events, such as proliferation and survival of LCs in the steady state.

Dicer-deficient Langerhans cells are migratory but display differential capacities to activate CD8^{+} versus CD4^{+} T cells

To analyze if Dicer$^{\Delta/\Delta}$ LCs displayed functional deficiencies before they undergo apoptosis, we next assessed the LC migratory capacity, a hallmark of LC biology. To this end, ear explants were cultured in medium containing the chemokine CCL21, and we analyzed by FACS cells migrating out of the explants. As shown in Fig. 6*A*, numbers of migratory epidermal LCs from Dicer$^{\Delta/\Delta}$ mice are strongly reduced compared with those from controls. This reduction highly correlated with the reduced number of LCs in the skin of Dicer$^{\Delta/\Delta}$ mice in steady-state conditions (Fig. 3*B*), implying that the Dicer-deficient LCs can be mobilized comparable to wild-type LCs upon a chemotactic stimulus. Culture of epidermal sheets is known to induce maturation of LCs, and a concomitant modulation of typical maturation markers on the surface of LCs can be observed (35). Consequently, wild-type LCs upregulated surface expression of MHC class I and class II molecules, CCR7, CD40, and CD86 upon exit of the epidermis (Fig. 6*B*). In marked contrast, Dicer-deficient LCs were unable to upregulate these molecules to the same extent (Fig. 6*B*), except for MHC class I and CCR7, which were modulated normally (Fig. 6*B*). Upregulation of CCR7, the chemokine receptor for CCL21, allowed Dicer$^{\Delta/\Delta}$ LCs to migrate out of the epidermal layers of the skin (Fig. 6*A*). Functional consequences of this maturation deficiency were observed in Ag presentation assays with T cells. When LCs were incubated with OVA protein overnight and cocultured with OVA-specific CD8^{+} OT-I or CD4^{+} OT-II cells, Dicer$^{wt/wt}$ LCs efficiently induced proliferation of OT-II cells, as measured by dilution of the CFSE dye (Fig. 6*C*). In contrast, Dicer$^{\Delta/\Delta}$ LCs could induce only a 2- to 3-fold lower percentage of divided T cells in these assays (Fig. 6*C*). Although Ag presentation via MHC class II was severely inhibited in absence of Dicer, we could not detect differences in priming of CD8^{+} OT-I T cells. These findings indicate that the uptake of exogenous protein Ag and its consecutive crosspresentation via MHC class I was not affected by Dicer deficiency (Fig. 6*C*) and are consistent with the intact upregulation of MHC class I by Dicer$^{\Delta/\Delta}$ LCs (Fig. 6*B*). Taken together, these data show that Dicer regulates the phenotypical and functional maturation of LCs and that absence of Dicer renders LCs defective to activate MHC class II-restricted CD4^{+} T cells.

A common method to test for cutaneous immune responses is the topical application of hapten Ag to the skin to induce contact hypersensitivity (CHS). It is thought that LCs play a central role in this skin-mediated immunity, although recent findings with a different LC ablation model generated controversial data (reviewed in Ref. 36). To test the function of Dicer$^{\Delta/\Delta}$ LCs in vivo, we sensitized and challenged the mice with DNFB. The Dicer$^{\Delta/\Delta}$ mice used for these tests were older than 9 wk and had already considerably lower numbers of LCs in their epidermis as compared with those in wild-type mice (Fig. 4*C*). Nevertheless, the measured ear swelling was similar in both groups (Fig. 6*D*). This might in part be due to the fact that DNFB can also be transported to the lymph nodes by dermal DCs, and therefore this test seems not to address specifically the function of LCs, as previously suggested (37). We therefore performed CHS with the hapten Oxazolone at a low dose previously described to selectively target LCs (38). However, also in this setting, no difference between Dicer$^{\Delta/\Delta}$ mice and Dicer$^{wt/wt}$ mice could be observed (Fig. 6*D*). Taken together, our data suggest that Dicer controls LC maturation and Ag presentation via MHC class II, but this deficiency does not directly cause deficient skin immunity as tested with CHS.

Discussion

In this study, we have shown that loss of Dicer leads to progressive ablation of LCs in the epidermis of the skin. As indicated by the almost complete absence of miR-223, the CD11c promoter-driven expression of Cre recombinase resulted in efficient disruption of *dicer1* in these cells. Dicer-deficient LCs showed lack of Birbeck granules, normally a typical attribute of LCs, furthermore, a disturbed expression of surface molecules and reduced Ag presentation capacities to CD4^{+} T cells, increased turnover, reduced half-lives, and increased rates of apoptosis.

A complex molecular network of signal transducers, transcription factors, and miRNAs is emerging controlling proliferation and apoptosis (39, 40). One of the best-studied miRNA families in this regard is the miR-17-92 cluster. Two studies identified Bim as a target for miRNAs of the miR-17-92 cluster and showed elevated levels of Bim protein in its absence (28, 41). Dicer-deficient LCs in our study expressed nearly 6-fold elevated mRNA levels of the

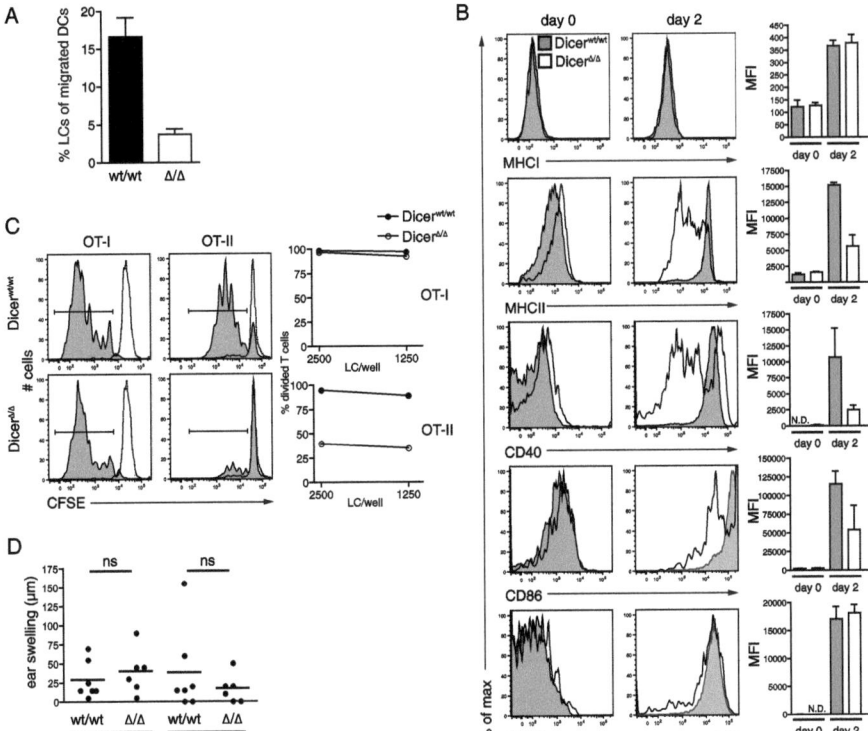

FIGURE 6. Dicer deficiency compromises maturation and function of LCs. *A*, DCs migrating out of ear explants from Dicer$^{wt/wt}$ and Dicer$^{\Delta/\Delta}$ animals in the presence of CCL21 were analyzed by FACS. Percentages of migrated Langerin$^+$CD103$^-$ LCs are displayed (mean ± SEM, $n = 6$; combined results from two independent experiments are shown). *B*, Epidermal sheets from Dicer$^{wt/wt}$ and Dicer$^{\Delta/\Delta}$ animals were digested to release LCs immediately (day 0) or cultured for two more days (day 2). Expression of surface markers on CD45$^+$Langerin$^+$ LCs was assessed using FACS. Shown are histogram overlays of Dicer$^{wt/wt}$ (gray) and Dicer$^{\Delta/\Delta}$ (open) LC staining intensities. Results are representative of two experiments ($n = 2$–6). *C*, Epidermal skin explants were cultured overnight in medium containing 0.25 mg/ml OVA protein and after washing further cultured until day 2. Emigrated LCs were enriched, and equal numbers of LCs were cocultured with CFSE-labeled OT-I or OT-II T cells. T cell proliferation was assessed by flow cytometry after 60–65 h. One representative experiment out of two independent experiments with a similar outcome is shown (OT-II, $n = 6$; OT-I, $n = 3$–4). Dashed line, control without LCs. The graph shows the percentage of T cells having proliferated with graded doses of LCs. *D*, Dicer$^{wt/wt}$ and Dicer$^{\Delta/\Delta}$ animals were sensitized with hapten (0.3% DNFB or 0.5% Oxazolone) on their backs. Five days later, ear swelling was elicited by challenging the mice with the same hapten on both sides of one ear. After 24 h, ear swelling was measured as the difference between ear thickness prior to and after challenge. Points represent individual mice ($n = 6$–7). N.D., not detectable.

proapoptotic Bim, indicating a potential involvement of this apoptosis pathway also in the disappearance of Dicer$^{\Delta/\Delta}$ LCs. One possibility for the lack of effect on spleen DC populations after *dicer1* deletion could be the appearance of DCs that escaped deletion of the *dicer1* alleles. However, in very young mice also, Dicer$^{\Delta/\Delta}$ LCs were present in the epidermis in normal numbers but decreased to almost complete absence with increasing age of the mice. This is in contrast to a previous report describing skin-repopulating LCs that escaped Cre-mediated deletion of *TGFβ1* and *TGFβRII* alleles (42), but such a phenomenon certainly depends on the type of promoter driving the Cre transgene.

However, although >90% of LCs do already express Cre and are RFP$^+$ at the age of 10 d, the initial seeding of the epidermis with LCs in Dicer$^{\Delta/\Delta}$ mice is normal. We therefore assume that at this time point Dicer has been inactivated recently, but effects are not yet visible. Similarly, >90% of DCs in spleen and lymph nodes were RFP$^+$ in CD11c-Cre-tdRFP-Dicer$^{fl/fl}$ mice (data not shown), and Dicer mRNA was undetectable in these cells, but they developed and functioned normally. The presence of substantial amounts of miRNAs at nearly undetectable levels of Dicer protein in spleen DCs could have several possible reasons. First, miRNAs can eventually be generated in the absence of Dicer. However, this seems unlikely because other enzymes could not replace Dicer functionally in ubiquitously Dicer-deficient mice, which are not viable (43). Also, Dicer is necessary to generate mature miRNAs in T cells, and its function was not redundant (24). Second, the highly reduced amounts of Dicer protein in spleen DCs are sufficient to generate mature miRNAs. Although it seems unlikely, we cannot generally rule out this possibility. Third, the half-life of miRNAs is too long and does not decrease beyond critical levels during the life span of Dicer$^{\Delta/\Delta}$ spleen DCs. A delay between the disappearance of Dicer protein and mature miRNAs might be considered as a reason for an absent phenotype in spleen and lymph node DCs as well as normal initial LC development. A recent study also described such a delay after Cre-mediated Dicer deletion early in embryogenesis. Careful examination of residual miRNA expression in affected tissues suggests a 3–10 d delay between elimination of Dicer and depletion of specific miRNAs (44). The mechanisms influencing miRNA stability are not yet fully understood, although multiple factors including cis- and trans-acting modifications and proteins regulating their half-lives have been discovered (45). Generally, miRNAs are considered to be stable molecules, with half-lives ranging from hours to days. Certain miRNAs, such as miRNA-208, have a half-life of >12 d (46). In contrast, BrdU studies determined the half-life of DCs in mesenteric lymph nodes and spleen to be 1.5–3 d and of DCs in cutaneous lymph nodes as 7–9 d (47, 48). Tissue-derived lymph node DCs had a half-life of 5–7 d as determined in parabiosis studies (49). It is therefore possible that the half-life of at least some mature miRNAs might be longer as compared with the half-life of spleen DCs. Therefore, Dicer$^{\Delta/\Delta}$ DCs of spleen and lymph nodes may contain functional miRNAs despite efficient deletion of Dicer. We detected a <50% decrease for some selected miRNAs, which could indicate that only a part of the spleen DC miRNA content might be due to de novo synthesis, whereas a large part could be residual molecules.

LCs gain cell size, and it is possible that this is due to the decrease of LC population density, leaving more "space" or "niches" available for each individual remaining LC. In Dicer$^{\Delta/\Delta}$ LCs, we could identify differential regulation of myc, which, when overexpressed, increases cell size during all stages of B cell differentiation (50). However, the loss of Dicer instead caused a strongly decreased expression of myc in LCs, and it is hardly conceivable that such a loss of expression may lead to the observed increased cell size of Dicer$^{\Delta/\Delta}$ LCs. Myc is also an important member of the cellular proliferation regulatory network and is part of a well-studied feedback loop involving the miR-17-92 cluster miRNAs and the critical cell cycle regulator E2F (39). Recently, other miRNAs have also been discovered that target Myc (51, 52). Both up- as well as downregulation of myc expression are linked to apoptosis, depending on the experimental system used (53, 54). Thus, the decreased levels of Myc mRNA in Dicer$^{\Delta/\Delta}$ LCs might also contribute to apoptosis. Also, the observed loss of TGFβRII expression might contribute to LC disappearance, because TGFβRII-deficient LCs are lost from epidermis (42). Although TGFβRII expression did not disappear completely in Dicer$^{\Delta/\Delta}$ LCs, the observed lower expression levels could be sufficient to cause survival disadvantages.

We found that CHS is not altered in Dicer$^{\Delta/\Delta}$ mice. Currently, the role of LCs in skin immunity is controversially discussed. With gene-targeted mice where LCs can be ablated by the use of diphtheria toxin, it was found that CHS responses were either reduced in one model (37) or completely unaffected (55) by LC ablation in a second model. With a constitutive LC ablation model, it was even found that CHS was increased, suggesting a regulatory function for LCs (56). Therefore, it is still unclear which roles LCs play in skin immunity (34). We used Dicer$^{\Delta/\Delta}$ mice that were older than 9 wk in our CHS assays. We could show that at this age the absence of Dicer had caused already a severe loss of LCs, with nearly no LCs left in the skin. Therefore, our results indicate that either the few remaining Dicer$^{\Delta/\Delta}$ LCs were sufficient to elicit hypersensitivity or, as suggested in other models (55), that LCs play no obvious role in CHS.

Our finding that Dicer-deficient LCs fail to upregulate maturation markers is in line with other reports showing implications for miRNAs at certain developmental stages (17). MiR-155, which is upregulated in human monocyte-derived DCs in response to LPS (19, 21), has been shown to play a role in DC maturation. However, miR-155–deficient DCs do not fail to upregulate maturation markers (18); therefore, other miRNAs lacking in Dicer$^{\Delta/\Delta}$ LCs are likely to be involved in the maturation process. Thus, the reduced Ag presentation capacity observed in Dicer$^{\Delta/\Delta}$ LCs might be a consequence of miRNA-dependent regulation of DC activation. The fact that only MHC class II-mediated presentation is inhibited, but not MHC class I cross-presentation, indicates that uptake of protein Ag might not be affected by loss of Dicer. Currently, we do not know which genes are targets of miRNA regulation in this context nor if the deficiency to upregulate MHC class II can be overcome by further signals, such as anti-CD40 or microbial stimuli. This issue is subject to further studies.

We also attempted to identify whether specific miRNAs contributed to the observed phenotype in Dicer-deficient LCs via correlation of Dicer-dependent gene expression changes with occurrence of specific miRNA seed motifs in the 3' untranslated region of affected genes, as has been reported for Dicer-deficient B cells (28). However, using the recently developed Sylamer software (57), we did not identify a specific miRNA responsible for the gene expression changes between wild-type and Dicer$^{\Delta/\Delta}$ LCs. The fact that reduced levels of many miRNAs affect simultaneously the expression of many genes, compounded by secondary effects of genes influencing each other's expression, might account for the failure to identify specific miRNAs. Specific deletion of individual miRNAs or miRNA clusters in LCs is required to gain more insight into the role of individual miRNAs in these processes.

Acknowledgments

We thank G. Hannon and B. Reizis for providing the Dicer^{flx} mice and the CD11c-Cre mice, respectively and D. Livingston for the anti-Dicer Ab. We thank C. Ried and U. Fazekas for expert technical assistance and A. Bol, W. Mertl, S. Heinzl, and M. Grdic for excellent animal care.

Disclosures

The authors have no financial conflicts of interest.

References

1. Shortman, K., and Y. J. Liu. 2002. Mouse and human dendritic cell subtypes. *Nat. Rev. Immunol.* 2: 151–161.
2. Onai, N., A. Obata-Onai, M. A. Schmid, T. Ohteki, D. Jarrossay, and M. G. Manz. 2007. Identification of clonogenic common Flt3+M-CSFR+ plasmacytoid and conventional dendritic cell progenitors in mouse bone marrow. *Nat. Immunol.* 8: 1207–1216.
3. Naik, S. H., P. Sathe, H. Y. Park, D. Metcalf, A. I. Proietto, A. Dakic, S. Carotta, M. O'Keeffe, M. Bahlo, A. Papenfuss, et al. 2007. Development of plasmacytoid and conventional dendritic cell subtypes from single precursor cells derived in vitro and in vivo. *Nat. Immunol.* 8: 1217–1226.
4. Shortman, K., and S. H. Naik. 2007. Steady-state and inflammatory dendritic-cell development. *Nat. Rev. Immunol.* 7: 19–30.
5. Ginhoux, F., M. P. Collin, M. Bogunovic, M. Abel, M. Leboeuf, J. Helft, J. Ochando, A. Kissenpfennig, B. Malissen, M. Grisotto, et al. 2007. Blood-derived dermal langerin+ dendritic cells survey the skin in the steady state. *J. Exp. Med.* 204: 3133–3146.
6. Bursch, L. S., L. Wang, B. Igyarto, A. Kissenpfennig, B. Malissen, D. H. Kaplan, and K. A. Hogquist. 2007. Identification of a novel population of Langerin+ dendritic cells. *J. Exp. Med.* 204: 3147–3156.
7. Poulin, L. F., S. Henri, B. de Bovis, E. Devilard, A. Kissenpfennig, and B. Malissen. 2007. The dermis contains langerin+ dendritic cells that develop and function independently of epidermal Langerhans cells. *J. Exp. Med.* 204: 3119–3131.
8. Ohnmacht, C., A. Pullner, S. B. King, I. Drexler, S. Meier, T. Brocker, and D. Voehringer. 2009. Constitutive ablation of dendritic cells breaks self-tolerance of CD4 T cells and results in spontaneous fatal autoimmunity. *J. Exp. Med.* 206: 549–559.

9. Jung, S., D. Unutmaz, P. Wong, G. Sano, K. De los Santos, T. Sparwasser, S. Wu, S. Vuthoori, K. Ko, F. Zavala, et al. 2002. In vivo depletion of CD11c(+) dendritic cells abrogates priming of CD8(+) T cells by exogenous cell-associated antigens. *Immunity* 17: 211–220.
10. He, Y., J. Zhang, C. Donahue, and L. D. Falo, Jr. 2006. Skin-derived dendritic cells induce potent CD8(+) T cell immunity in recombinant lentivector-mediated genetic immunization. *Immunity* 24: 643–656.
11. Allan, R. S., C. M. Smith, G. T. Belz, A. L. van Lint, L. M. Wakim, W. R. Heath, and F. R. Carbone. 2003. Epidermal viral immunity induced by CD8alpha+ dendritic cells but not by Langerhans cells. *Science* 301: 1925–1928.
12. Dudziak, D., A. O. Kamphorst, G. F. Heidkamp, V. R. Buchholz, C. Trumpfheller, S. Yamazaki, C. Cheong, K. Liu, H. W. Lee, C. G. Park, et al. 2007. Differential antigen processing by dendritic cell subsets in vivo. *Science* 315: 107–111.
13. Luckashenak, N., S. Schroeder, K. Endt, D. Schmitt, K. Mahnke, M. F. Bachmann, P. Marconi, C. A. Deeg, and T. Brocker. 2008. Constitutive crosspresentation of tissue antigens by dendritic cells controls CD8+ T cell tolerance in vivo. *Immunity* 28: 521–532.
14. Wu, L., and Y. J. Liu. 2007. Development of dendritic-cell lineages. *Immunity* 26: 741–750.
15. Hildner, K., B. T. Edelson, W. E. Purtha, M. Diamond, H. Matsushita, M. Kohyama, B. Calderon, D. U. Schraml, E. R. Unanue, M. S. Diamond, et al. 2008. Batf3 deficiency reveals a critical role for CD8alpha+ dendritic cells in cytotoxic T cell immunity. *Science* 322: 1097–1100.
16. Bartel, D. P. 2009. MicroRNAs: target recognition and regulatory functions. *Cell* 136: 215–233.
17. Lindsay, M. A. 2008. microRNAs and the immune response. *Trends Immunol.* 29: 343–351.
18. Rodriguez, A., E. Vigorito, S. Clare, M. V. Warren, P. Couttet, D. R. Soond, S. van Dongen, R. J. Grocock, P. P. Das, E. A. Miska, et al. 2007. Requirement of bic/microRNA-155 for normal immune function. *Science* 316: 608–611.
19. Ceppi, M., P. M. Pereira, I. Dunand-Sauthier, E. Barras, W. Reith, M. A. Santos, and P. Pierre. 2009. MicroRNA-155 modulates the interleukin-1 signaling pathway in activated human monocyte-derived dendritic cells. *Proc. Natl. Acad. Sci. USA* 106: 2735–2740.
20. Hashimi, S. T., J. A. Fulcher, M. H. Chang, L. Gov, S. Wang, and B. Lee. 2009. MicroRNA profiling identifies miR-34a and miR-21 and their target genes JAG1 and WNT1 in the coordinate regulation of dendritic cell differentiation. *Blood* 114: 404–414.
21. Martinez-Nunez, R. T., F. Louafi, P. S. Friedmann, and T. Sanchez-Elsner. 2009. MicroRNA-155 modulates the pathogen binding ability of dendritic cells (DCs) by down-regulation of DC-specific intercellular adhesion molecule-3 grabbing non-integrin (DC-SIGN). *J. Biol. Chem.* 284: 16334–16342.
22. Cobb, B. S., A. Hertweck, J. Smith, E. O'Connor, D. Graf, T. Cook, S. T. Smale, S. Sakaguchi, F. J. Livesey, A. G. Fisher, and M. Merkenschlager. 2006. A role for Dicer in immune regulation. *J. Exp. Med.* 203: 2519–2527.
23. Cobb, B. S., T. B. Nesterova, E. Thompson, A. Hertweck, E. O'Connor, J. Godwin, C. B. Wilson, N. Brockdorff, A. G. Fisher, S. T. Smale, and M. Merkenschlager. 2005. T cell lineage choice and differentiation in the absence of the RNase III enzyme Dicer. *J. Exp. Med.* 201: 1367–1373.
24. Muljo, S. A., K. M. Ansel, C. Kanellopoulou, D. M. Livingston, A. Rao, and K. Rajewsky. 2005. Aberrant T cell differentiation in the absence of Dicer. *J. Exp. Med.* 202: 261–269.
25. Chong, M. M., J. P. Rasmussen, A. Y. Rudensky, A. Y. Rudensky, and D. R. Littman. 2008. The RNAseIII enzyme Drosha is critical in T cells for preventing lethal inflammatory disease. *J. Exp. Med.* 205: 2005–2017.
26. Cobb, B. S., A. L. F. Lu, D. O'Carroll, A. Tarakhovsky, and A. Y. Rudensky. 2008. Dicer-dependent microRNA pathway safeguards regulatory T cell function. *J. Exp. Med.* 205: 1993–2004.
27. Zhou, X., L. T. Jeker, B. T. Fife, S. Zhu, M. S. Anderson, M. T. McManus, and J. A. Bluestone. 2008. Selective miRNA disruption in T reg cells leads to uncontrolled autoimmunity. *J. Exp. Med.* 205: 1983–1991.
28. Koralov, S. B., S. A. Muljo, G. R. Galler, A. York, T. Chakraborty, C. Kanellopoulou, K. Jensen, B. S. Cobb, M. Merkenschlager, N. Rajewsky, and K. Rajewsky. 2008. Dicer ablation affects antibody diversity and cell survival in the B lymphocyte lineage. *Cell* 132: 860–874.
29. Andl, T., E. P. Murchison, F. Liu, Y. Zhang, M. Yunta-Gonzalez, J. W. Tobias, C. D. Andl, J. T. Seykora, G. J. Hannon, and S. E. Millar. 2006. The miRNA-processing enzyme dicer is essential for the morphogenesis and maintenance of hair follicles. *Curr. Biol.* 16: 1041–1049.
30. Caton, M. L., M. R. Smith-Raska, and B. Reizis. 2007. Notch-RBP-J signaling controls the homeostasis of CD8- dendritic cells in the spleen. *J. Exp. Med.* 204: 1653–1664.
31. Luche, H., O. Weber, T. Nageswara Rao, C. Blum, and H. J. Fehling. 2007. Faithful activation of an extra-bright red fluorescent protein in "knock-in" Cre-reporter mice ideally suited for lineage tracing studies. *Eur. J. Immunol.* 37: 43–53.
32. Lauterbach, H., C. Ried, A. L. Epstein, P. Marconi, and T. Brocker. 2005. Reduced immune responses after vaccination with a recombinant herpes simplex virus type 1 vector in the presence of antiviral immunity. *J. Gen. Virol.* 86: 2401–2410.
33. Landgraf, P., M. Rusu, R. Sheridan, N. Iovino, A. Aravin, S. Pfeffer, A. Rice, A. O. Kamphorst, M. Landthaler, et al. 2007. A mammalian microRNA expression atlas based on small RNA library sequencing. *Cell* 129: 1401–1414.
34. Merad, M., F. Ginhoux, and M. Collin. 2008. Origin, homeostasis and function of Langerhans cells and other langerin-expressing dendritic cells. *Nat. Rev. Immunol.* 8: 935–947.
35. Ortner, U., K. Inaba, F. Koch, M. Heine, M. Miwa, G. Schuler, and N. Romani. 1996. An improved isolation method for murine migratory cutaneous dendritic cells. *J. Immunol. Methods* 193: 71–79.
36. Kaplan, D. H., A. Kissenpfennig, and B. E. Clausen. 2008. Insights into Langerhans cell function from Langerhans cell ablation models. *Eur. J. Immunol.* 38: 2369–2376.
37. Bennett, C. L., E. van Rijn, S. Jung, K. Inaba, R. M. Steinman, M. L. Kapsenberg, and B. E. Clausen. 2005. Inducible ablation of mouse Langerhans cells diminishes but fails to abrogate contact hypersensitivity. *J. Cell Biol.* 169: 569–576.
38. Bennett, C. L., M. Noordegraaf, C. A. Martina, and B. E. Clausen. 2007. Langerhans cells are required for efficient presentation of topically applied hapten to T cells. *J. Immunol.* 179: 6830–6835.
39. O'Donnell, K. A., E. A. Wentzel, K. I. Zeller, C. V. Dang, and J. T. Mendell. 2005. c-Myc-regulated microRNAs modulate E2F1 expression. *Nature* 435: 839–843.
40. Xiao, C., and K. Rajewsky. 2009. MicroRNA control in the immune system: basic principles. *Cell* 136: 26–36.
41. Ventura, A., A. G. Young, M. M. Winslow, L. Lintault, A. Meissner, S. J. Erkeland, J. Newman, R. T. Bronson, D. M. Crowley, J. R. Stone, et al. 2008. Targeted deletion reveals essential and overlapping functions for the miR-17 through 92 family of miRNA clusters. *Cell* 132: 875–886.
42. Kaplan, D. H., M. O. Li, M. C. Jenison, W. D. Shlomchik, R. A. Flavell, and M. J. Shlomchik. 2007. Autocrine/paracrine TGFbeta1 is required for the development of epidermal Langerhans cells. *J. Exp. Med.* 204: 2545–2552.
43. Bernstein, E., S. Y. Kim, M. A. Carmell, E. P. Murchison, H. Alcorn, M. Z. Li, A. A. Mills, S. J. Elledge, K. V. Anderson, and G. J. Hannon. 2003. Dicer is essential for mouse development. *Nat. Genet.* 35: 215–217.
44. Soukup, G. A., B. Fritzsch, M. L. Pierce, D. M. Weston, I. Jahan, M. T. McManus, and B. D. Harfe. 2009. Residual microRNA expression dictates the extent of inner ear development in conditional Dicer knockout mice. *Dev. Biol.* 328: 328–341.
45. Kai, Z. S., and A. E. Pasquinelli. 2010. MicroRNA assassins: factors that regulate the disappearance of miRNAs. *Nat. Struct. Mol. Biol.* 17: 5–10.
46. van Rooij, E., L. B. Sutherland, X. Qi, J. A. Richardson, J. Hill, and E. N. Olson. 2007. Control of stress-dependent cardiac growth and gene expression by a microRNA. *Science* 316: 575–579.
47. Kamath, A. T., J. Pooley, M. A. O'Keeffe, D. Vremec, Y. Zhan, A. M. Lew, A. D'Amico, L. Wu, D. F. Tough, and K. Shortman. 2000. The development, maturation, and turnover rate of mouse spleen dendritic cell populations. *J. Immunol.* 165: 6762–6770.
48. Kamath, A. T., S. Henri, F. Battye, D. F. Tough, and K. Shortman. 2002. Developmental kinetics and lifespan of dendritic cells in mouse lymphoid organs. *Blood* 100: 1734–1741.
49. Liu, K., C. Waskow, X. Liu, K. Yao, J. Hoh, and M. Nussenzweig. 2007. Origin of dendritic cells in peripheral lymphoid organs of mice. *Nat. Immunol.* 8: 578–583.
50. Iritani, B. M., and R. N. Eisenman. 1999. c-Myc enhances protein synthesis and cell size during B lymphocyte development. *Proc. Natl. Acad. Sci. USA* 96: 13180–13185.
51. Sachdeva, M., S. Zhu, F. Wu, H. Wu, V. Walia, S. Kumar, R. Elble, K. Watabe, and Y. Y. Mo. 2009. p53 represses c-Myc through induction of the tumor suppressor miR-145. *Proc. Natl. Acad. Sci. USA* 106: 3207–3212.
52. Lal, A., F. Navarro, C. A. Maher, L. E. Maliszewski, N. Yan, E. O'Day, D. Chowdhry, D. M. Dykxhoorn, P. Tsai, O. Hofmann, et al. 2009. miR-24 Inhibits cell proliferation by targeting E2F2, MYC, and other cell-cycle genes via "seedless" 3'UTR microRNA recognition elements. *Mol. Cell* 35: 610–625.
53. Thompson, E. B. 1998. The many roles of c-Myc in apoptosis. *Annu. Rev. Physiol.* 60: 575–600.
54. Dubois, N. C., C. Adolphe, A. Ehninger, R. A. Wang, E. J. Robertson, and A. Trumpp. 2008. Placental rescue reveals a sole requirement for c-Myc in embryonic erythroblast survival and hematopoietic stem cell function. *Development* 135: 2455–2465.
55. Kissenpfennig, A., S. Henri, B. Dubois, C. Laplace-Builhé, P. Perrin, N. Romani, C. H. Tripp, P. Douillard, L. Leserman, D. Kaiserlian, et al. 2005. Dynamics and function of Langerhans cells in vivo: dermal dendritic cells colonize lymph node areas distinct from slower migrating Langerhans cells. *Immunity* 22: 643–654.
56. Kaplan, D. H., M. C. Jenison, S. Saeland, W. D. Shlomchik, and M. J. Shlomchik. 2005. Epidermal langerhans cell-deficient mice develop enhanced contact hypersensitivity. *Immunity* 23: 611–620.
57. van Dongen, S., C. Abreu-Goodger, and A. J. Enright. 2008. Detecting microRNA binding and siRNA off-target effects from expression data. *Nat. Methods* 5: 1023–1025.

Supplemental Figure 1:

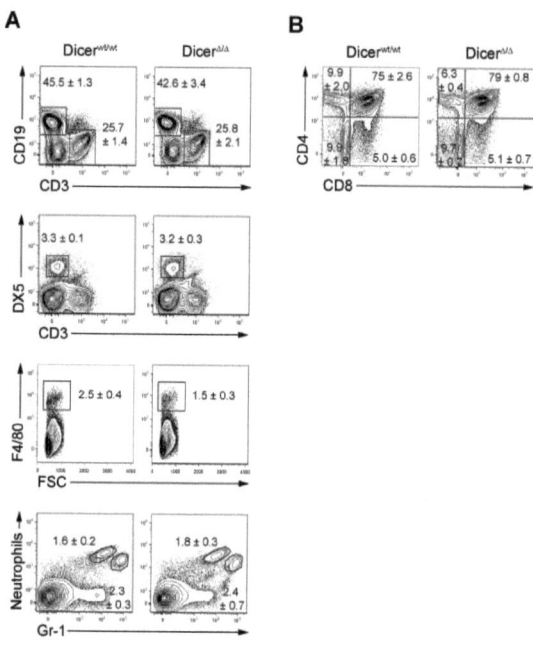

Figure S1. *CD11c-Cre-mediated deletion of Dicer has no effect on non-DC leukocyte populations.* FACS analysis of B cells, T cells, NK cells, macrophages and neutrophils in the spleen (**A**) and T cells in the thymus (**B**). Percentages of cells that fall into each gate (mean ± SEM, n=6) are indicated. All populations depicted were gated on live cells; for analysis of neutrophils F4/80⁻ cells were gated.

7 Discussion

Cells undergoing developmental transformation or terminal differentiation dramatically alter their mRNA profile. Such changes in cellular identity are predominantly driven by transcription factors. However, the interaction of transcription factors with miRNAs, which, according to Kosik et al., can 'canalize' gene expression by operating in feedback or feedforward loops, may ultimately stabilize or promote the new cellular identity (reviewed in Kosik, 2010).

DCs require transcription factors to guide their differentiation from hematopoietic precursors via various intermediates into the different DC subsets (reviewed in Zenke and Hieronymus, 2006). When immature DCs in the steady-state encounter pathogens, it is vital that they can quickly respond to the threat by becoming mature, in order to initiate protective adaptive immune responses. Research in the past few years has shown that transition states, such as development and terminal differentiation, are cellular conditions in which miRNAs are typically involved. This has been well documented for other immune cells, in particular for T and B lymphocytes. For example, the survival of lymphocyte progenitors relies on the expression of miR-17~92, while their terminal differentiation in response to activating signals is regulated by miR-155 (reviewed in O'Connell et al., 2010b).

Given the various cell types and subsets thereof that arise from hematopoietic stem cells, it is conceivable that transcription factor-controlled DC lineage and subset fate decisions may be reinforced by miRNAs. As DCs need to respond to and integrate environmental signals rapidly, the concept of miRNA control of the maturation process appears reasonable. In the following discussion, I provide evidence that both DC differentiation and maturation is critically dependent on miRNA regulation.

7.1 miRNAs regulate DC differentiation

We analyzed the miRNA expression profiles in different DC subsets (Publication I) and identified several miRNAs that were differentially expressed in Flt3L BM-derived cDCs and pDCs. Moreover, some selected miRNAs were differently expressed in inflammatory DCs, which were generated in GM-CSF BM cultures, as compared to Flt3L cDCs. We have further demonstrated that miRNAs can influence DC lineage fate decisions, as knockdown of miR-221 or miR-222 during DC differentiation altered the ratio of pDCs/cDCs towards a higher pDC frequency *in vitro*.

For our miRNA analysis we used cDCs and pDCs generated in Flt3L BM cultures as they can be generated in large numbers and closely resemble their *in vivo* counterparts. They share properties such as surface marker expression (apart from CD4 and CD8 expression), TLR and chemokine receptor expression, cytokine production upon stimulation and, importantly, transcription factor expression and dependence for differentiation (Naik et al., 2005). Given these similarities, we assume that miRNAs regulate differentiation and function of *in vivo* and *in vitro* generated DCs analogously.

An increasing number of reports indicate that miRNAs play a role in the differentiation of DCs, with certain miRNAs being identified as having have key roles in either DC lineage differentiation or DC subset differentiation, many of which are represented in our miRNA array data.

7.1.1 miRNAs regulate DC lineage and subset differentiation by targeting Notch/Wnt signaling

As described in section 4.1.2, CDPs develop from MDPs in the BM and split then into pDC and cDC lineages. Notch signaling appears to direct the differentiation from MDPs into CDPs (Ohishi, 2001). Cooperative action between the Notch and the downstream Wnt signaling pathway is thought to be required for steering CDPs into cDC direction. In contrast, directing CDPs towards pDC differentiation requires Notch signaling, but may be blocked by Wnt signaling (Zhou et al., 2009a). Thus, Wnt signaling might determine the direction of DC differentiation (reviewed in Cheng et al., 2010).

In our study, we found that miR-21 was ~7-fold more highly expressed in cDCs than in pDCs, and was similarly highly expressed in cDCs and GM-CSF DCs (Publication I, Table 1 and Fig. 4). Interestingly, Hashimi et al. showed a requirement for miR-21-induced repression of the Notch ligand

Jagged-1 in the terminal differentiation of human monocyte-derived DCs (Hashimi et al., 2009). pDCs, however, do not express Jagged-1, but express Notch ligands of the Delta-like (Dll) family instead (Kassner et al., 2010). Remarkably, Dll-1 is predicted by TargetScan5.1 to contain binding sites for several miRNAs, miR-15b, miR-103 and miR-107, all of which are overrepresented in pDCs compared to cDCs (unpublished observation and Publication I, Table 1). Thus, in order to allow for differentiation of DC subsets, miR-21 expression might be relevant for cDC differentiation, while, analogously, miR-15b, miR-103 and miR-107 expression might be relevant for pDC differentiation. The different requirements are possibly reflected in the respective cDC and pDC miRNA expression patterns. Besides miR-21, Hashimi et al. also found miR-34a to be important for DC differentiation from monocytes. miR-34a acts through targeting of Wnt-1 (Hashimi et al., 2009), which is believed to be downstream of Notch (reviewed in Cheng et al., 2010). However, we did not observe differential expression of miR-34a between pDCs and cDCs, suggesting that miR-34a-mediated blocking of Wnt-1 is generally important for DC lineage differentiation, but not for DC subset specification. Nevertheless, more miRNAs directing DC lineage differentiation by targeting the Notch and Wnt pathways might be identified. Supporting this notion, the Notch signaling pathway has been shown to be a major target of miRNA regulation in Drosophila (Lai et al., 2005) and there is evidence of miRNA targeting of a Notch ligand in mice and humans (Ivey et al., 2008).

7.1.2 miRNAs influence DC subset differentiation by controlling the expression of transcription factors

Evidence for miRNA regulation of DC differentiation comes from a recent study by Lu et al., who profiled miRNA expression in human monocytes, immature monocyte-derived GM-CSF DCs and mature monocyte-derived GM-CSF DCs (Lu et al., 2011). They found miR-221 to be upregulated in immature DCs, which was followed by downregulation of miR-221 expression upon DC maturation. In addition, they provide evidence that the cell cycle inhibitor $p27^{kip1}$ is a target of miR-221 and needs to be suppressed during DC differentiation to prevent apoptosis. This suggests a role for miR-221 during DC development, while DC maturation does not seem to depend on miR-221 expression. Together with our finding that the clustered miRNAs miR-221 and miR-222 were more strongly expressed in cDCs than in pDCs (Publication I, Table 1 and Fig. 1B), these results indicate that miR-221 and miR-222 have a function in promoting cDC and GM-CSF DC differentiation, but may be irrelevant for differentiation into pDCs. Indeed, miR-221 action might even have an inhibitory effect on pDC differentiation, as we have shown that knockdown of miR-221 or miR-222 positively regulates pDC differentiation (Publication I, Fig. 3D).

Whether the upregulation of a single target gene accounts for the increased pDC/cDC ratio in the miR-221/222 knockdown DC cultures, or whether the targeting of multiple genes leads to the observed phenotype, remains to be tested. However, we have detected three miR-221/222 candidates that are likely to influence pDC versus cDC differentiation: the cell cycle inhibitor $p27^{kip1}$, the cytokine receptor c-kit, and the transcription factor E2-2.

(i) The cell cycle inhibitor $p27^{kip1}$ is a confirmed target of miR-221/222 and its function in cDC differentiation has been validated (Lu et al., 2011). We observed similar expression levels of $p27^{kip1}$ in Flt3L pDCs and cDCs (unpublished observations). Derepression of $p27^{kip1}$ by knockdown of miR-221/222 in Flt3L DC cultures might stall cDC precursor proliferation while not affecting pDC precursors, thus leading to a higher pDC/cDC ratio.

(ii) *Kit*, encoding for the stem cell factor receptor c-kit, is expressed on a common pDC and cDC precursor (Naik et al., 2007; Onai et al., 2007) and has been identified as a target of miR-221/222 in hematopoietic stem cells as well as in endothelial cells (Felli, 2005; Poliseno et al., 2006). Signaling via c-kit is known to promote proliferation and differentiation of hematopoietic stem cells (Kent et al., 2008). Enhanced signaling via c-kit might promote pDC precursor differentiation more strongly than cDC precursor differentiation. However, Fujita et al. have shown that an activating mutation in the c-kit signaling domain causes a decrease in the pDC/cDC ratio (Fujita et al., 2011). Thus, the influence of increased c-kit expression on pDC and cDC differentiation remains unclear.

(iii) When we probed TargetScan for potential miR-221/222 target genes, *tcf4*, encoding for E2-2, was one of the revealed potential targets. Interestingly, E2-2 is known as the master transcription factor for pDC lineage specification, controlling the pDC-specific gene expression program (Cisse et al., 2008; Ghosh et al., 2010). E2-2 is highly expressed in splenic pDCs but not in splenic cDCs (Cisse et al., 2008). Our analysis of E2-2 expression in *in vitro* generated pDCs and cDCs shows a similar preferential expression in pDCs (Publication I, Fig. 1B). An increased pDC frequency would be the expected consequence of released E2-2 repression by miR-221/222 knockdown, and exactly this was observed in our study (Publication I, Fig. 3D). Inversely, it has been shown that deletion of E2-2 in pDCs results in the acquisition of a cDC phenotype displaying cDC properties (Ghosh et al., 2010).

Thus, while it is not yet clear which mechanism causes the skewed pDC/cDC ratios in miR-221/222 knockdown DC cultures, it seems unlikely that the exclusive regulation of one gene accounts for the observed shift. Rather, as is typical for miRNAs, a transcriptional network that is coordinately influenced by miR-221/222 expression is likely to determine DC lineage specification.

pDC versus cDC lineage specification is not only regulated by the pDC transcription factor E2-2, but also by the E2-2 antagonist Id2. Id2 overexpression inhibits pDC differentiation (Spits et al., 2000), whereas Id2-deficient mice have increased pDC numbers and reduced numbers of $CD8\alpha^+$ DCs and LCs (Hacker et al., 2003). Although the antagonistic actions of E2-2 and Id2 probably play a dominant role in DC lineage commitment, it might be that, in accordance with a hypothesis by Wu et al. (Wu et al., 2009), miRNAs act to reduce transcriptional 'noise' and thus stabilize transcription factor-induced commitment, or tune the transcription factor expression profile and thus shift differentiation towards either pDC or cDC direction. The latter is likely to have occurred in our knockdown studies, where we see a shifted lineage commitment.

We have hypothesized that high miR-221/222 expression in cDCs stabilizes the cDC cell fate by suppressing E2-2. Conversely, could there be miRNAs that reinforce pDC lineage commitment? Using TargetScan, we analyzed whether Id2 contains miRNA binding sites and found that miR-19a and -19b are predicted to target Id2. Indeed, miR-19b is amongst the most differentially expressed miRNAs between cDCs and pDCs (Publication I, Table 1), and its expression is 3-fold higher in pDCs than in cDCs. Thus, miR-19b might function to control Id2 expression in pDC. Whether miR-221/222 and miR-19b represent a pair of antagonistic cell fate regulators that act on CDPs before lineage bifurcation, similar to the regulatory role that has been suggested for Id2 and E2-2 (reviewed in Reizis et al., 2010), is an intriguing question currently being studied.

Interestingly, our array data demonstrate that several members of the miR-17~92 and the homologous miR-106a~363 cluster are more highly expressed in pDCs than in cDCs. miR-19b, -18a and -20a belong to this set of clusters, as well as miR-106a (Publication I, Table 1). The miR-17~92 cluster has emerged as a critical regulator of cellular development (Mendell, 2008), and has been implicated in the suppression of TGF-β signaling in tumors (Petrocca et al., 2008). Notably, Felker et al. have recently demonstrated that TGF-β1 functions as a critical factor at the bifurcation of pDC versus cDC differentiation and specifically drives cDC subset specification by inducing Id2 (Felker et al., 2010). Therefore it would be interesting to analyze whether pDC differentiation requires enhanced expression of miR-17~92 to suppress TGF-β signaling. Furthermore, it has been shown that miR-17~92 miRNAs are highly expressed especially in lymphocyte progenitors (Xiao et al., 2008). The similarity between miRNAs expressed in pDC and lymphocytes is not restricted to the expression of miR-17~92 miRNAs. Our cluster analysis based on cDC, pDC and $CD4^+$ T cell miRNA expression patterns clearly shows that pDCs and T cells are more closely related than pDCs and cDCs in terms of miRNA expression (Publication I, Fig. 2B). It has been suggested that expression of E2-2 in the steady-state is responsible for the expression of key transcription factors in pDCs, such as Spib and Irf7, and imparts pDCs with lymphocyte-like features (section 4.1.1.3). Whether the miRNA expression pattern in pDCs is also determined by E2-2 expression is not yet known. As activated pDCs undergo cell fate conversion to cDCs (Liou et al., 2008), and this is accompanied by a several-fold reduction of E2-2 expression, it would be interesting to analyze the miRNA profile of activated converted pDCs and compare it with the profile of activated cDCs and lymphocytes, to see whether there is an adaptation of the miRNA patterns in activated pDCs and cDCs.

Some other miRNAs are noteworthy in respect to DC subset differentiation: the cluster miR-23a/24/27a and miR-223. We have revealed that miR-23a, -24, -27a and miR-223 are 2- and 3-fold more highly expressed in cDCs than in pDCs, respectively, while being similarly highly expressed in cDCs and GM-CSF DCs (Publication I, Table 1). It has been shown recently that the transcription factor PU.1 induces transcription of the clustered miRNAs miR-23a, -24, and -27a (Kong et al., 2010). Expression of miR-223 might also be driven by PU.1 (Fukao et al., 2007). PU.1 controls DC differentiation by inducing the expressing of Flt3 and GM-CSF receptors on hematopoietic progenitors (Carotta et al., 2010; Zhang et al., 1996). While being highly expressed on CDPs and cDCs, PU.1 is downregulated at the onset of pDC differentiation (Carotta et al., 2010). Downregulation of PU.1 would therefore be expected to lead to reduced levels of miR-23a, -24, and -27a upon pDC differentiation, and indeed this was observed in our data. Downregulation of PU.1 and in consequence miR-23a/24/27a might be required to stabilize the lymphoid-like gene expression in pDCs. This assumption is underscored by a report by Kong et al., who have demonstrated for miR-23a/24/27a suppression of lymphopoiesis, but positive regulation of myelopoiesis (Kong et al., 2010). miR-223, which is a hematopoietic lineage-specific miRNA, has also been implicated in promoting myeloid lineage development (Johnnidis et al., 2008). Thus, our data suggest that miRNAs involved in myeloid versus lymphoid lineage fate decisions also function at the cDC versus pDC lineage bifurcation. This is in support of the notion that pDCs actively regulate lineage-specific gene expression programs to "antagonize the 'default' cDC cell fate" (Ghosh et al., 2010).

Our miRNA expression analysis also indicates that cDCs express more miR 146a than pDCs. miR 146a is known to dampen inflammatory responses by acting in a negative feedback loop on TLR and NF-κB signaling (section 4.3.3). Jurkin et al. have reported differential expression of miR-146a in human *in vitro* generated myeloid DC subsets, and in particular demonstrated ~6-fold increased miR-146a expression levels in LCs as compared to GM-CSF DCs, which was driven by elevated levels of PU.1 in LCs (Jurkin et al., 2010). They concluded that constitutively high levels of miR-

146a might prevent inappropriate LC activation by commensal bacteria. Translated to our findings, this might indicate a higher activation threshold in cDCs and, inversely, an increased susceptibility to activating signals in pDCs. To our knowledge, this has not yet been demonstrated. However, Rettig et al. have shown that monocytes have a higher activation threshold than pDCs (Rettig et al., 2010). Whether GM-CSF DCs have a similar miR-146-regulated threshold as cDCs is not yet clear. Our analysis showed higher expression of miR-146b but lower expression of miR-146a in GM-CSF DCs than in cDCs, and vice versa (Publication I, Fig. 4). Differential induction of these miRNA isoforms is probably pathway-dependent, and the induction of miR-146b is known to be NF-κB-independent (Perry et al., 2009), which could provide an explanation for the pro-inflammatory nature of GM-CSF DCs. Thus, miR-146 might be another example of a miRNA having a 'rheostat' function, similar to that reported for miR-181a, which sets the threshold in T cell priming (Li et al., 2007).

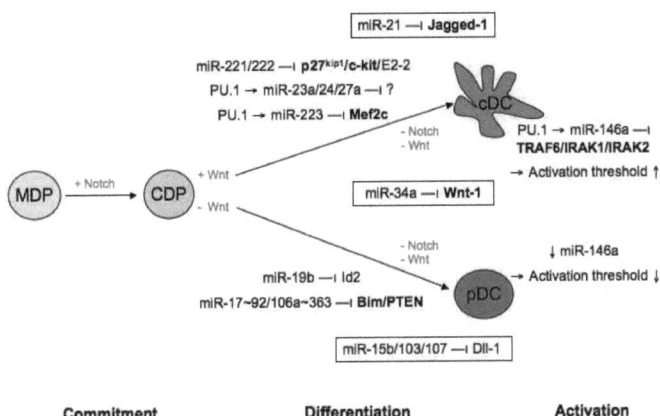

Figure 1: Possible miRNA regulation mechanisms directing or stabilizing cDC and pDC subset fates. Differentially expressed miRNAs between cDCs and pDCs may regulate DC differentiation and activation by targeting indicated molecules. For details see text. Verified miRNA targets are indicated in bold. miRNAs regulating Wnt/Notch ligands are framed. +/- Notch/Wnt, Notch/Wnt signaling required/inhibitory.

Taken together, the analysis of our data demonstrates that, although transcription factors are probably the dominant players in determining DC lineage or subset commitment, miRNAs may be critically involved in cell fate decisions by modulating the expression of transcription factors. Moreover, as pDCs and cDCs overexpress miRNAs that seemingly oppose the respective alternative cell fates, miRNAs might serve to stabilize cell fate decisions. Although mostly speculative, the plethora of plausible mechanisms I have presented here (summarized in Fig. 1) suggests that there is a substantial contribution of miRNAs in cell fate decisions.

7.2 miRNAs regulate DC survival and function

By crossing Dicer$^{fl/fl}$ mice with CD11c-Cre mice, we sought to generate Dicer$^{\Delta/\Delta}$ mice that specifically lack miRNAs in DCs (Publication II). Due to our assumption that deletion of Dicer would lead to a complete loss of mature miRNAs in DCs and hence to perturbed DC development, we expected the mice to have a compromised immune system, resulting for example in an autoimmune disease or a myeloid proliferative disorder, as has been observed in DC-deficient mice (Birnberg et al., 2008; Ohnmacht et al., 2009). However, this was not the case. Despite the absence of Dicer protein in splenic cDCs, we observed only slightly reduced miRNA levels in splenic and LN cDCs (Publication II, Fig. 1C and D), which is likely due to a several day delay between the loss of Dicer and the disappearance of miRNAs (Soukup et al., 2009). The average miRNA half-life has recently been reported to be ~5 days (Gantier et al., 2011). As DCs (excluding the LC subset) are short-lived cells, we suppose that this delay exceeds the DC life-span, and in consequence DC development is not affected (Publication II, Fig. 1E). Moreover, our Dicer$^{\Delta/\Delta}$ mice are immunocompetent and develop normal T and B cell responses (Publication II, Fig. 2), indicating that DCs are functionally intact. In contrast, it has been shown that the development of Dicer-deficient T and B cells is impaired (Koralov et al., 2008; Muljo, 2005), but T and B cells are thought to have a half-life of several weeks (Forster and Rajewsky, 1990; von Boehmer and Hafen, 1993), and therefore are more likely to completely lose miRNAs after Dicer ablation. Strikingly, however, we have demonstrated that LCs, which similarly to T and B cells have a half-life of several weeks in wild type mice, are severely affected in Dicer$^{\Delta/\Delta}$ mice (Publication II, Fig. 3D). We observed defective homeostasis of miRNA-deficient LCs resulting from increased turnover and apoptosis (Publication II, Fig. 5B), disturbed expression of surface receptors (Publication II, Fig. 3E), and, importantly, an inability to completely undergo the maturation process, which becomes apparent in the imperfect upregulation of costimulatory and MHC II molecules (Publication II, Fig. 6B). Presumably this defective maturation causes an inability to efficiently prime CD4$^+$ T cells (Publication II, Fig. 6C). Thus, miRNAs are not only important for the survival of LCs, but also for their hallmark function, the efficient antigen presentation and costimulation of T cells, that leads to the induction of adaptive responses.

7.2.1 miRNAs regulate DC turnover and apoptosis

With the exception of LCs, which live up to several weeks (section 4.1.1.2), DCs are short-lived cells with a half-half of several days (reviewed in Merad and Manz, 2009). Chen et al. have demonstrated that the ratio between pro- and anti-apoptotic molecules controls induction of apoptosis and consequently DC lifespan (Chen et al., 2007a). Regulation of DC lifespan is thought to be important to maintain tolerance, as experimentally prolonged DC survival leads to the onset of autoimmunity (Chen, 2006).

We have shown that the epidermal LC network forms properly in young Dicer$^{\Delta/\Delta}$ mice, but that LCs are then progressively lost from the epidermis, leading to their almost complete absence at the age of several months (Publication II, Fig. 4B and C). Chorro et al. have demonstrated that LCs seed the epidermis before birth and start to express CD11c around d0 to d2 after birth. LCs then undergo massive proliferation, which is completed after the first week of life (Chorro et al., 2009). With respect to our conditional knockout mice this means that Dicer$^{\Delta/\Delta}$ LCs start to lose Dicer protein shortly after birth, when CD11c is expressed. At this time, with their existing pool of miRNAs, LCs are able to proliferate and establish the LC network, which we could see in 10 day old mice (Publication II, Fig. 4B). However, LCs are not able to maintain the production of mature miRNAs due to ablation of Dicer, ultimately leading to loss of LCs. We observed an ~7-fold increased apoptosis rate in Dicer$^{\Delta/\Delta}$ LCs (Publication II, Fig. 5B), which is very likely to account for their loss. In support of this, we found a pro-apoptotic gene expression signature in Dicer$^{\Delta/\Delta}$ LCs and verified the upregulation of the pro-apoptotic molecule Bim in Dicer$^{\Delta/\Delta}$ LCs (Publication II, Fig. 5A and text). Moreover, Dicer$^{\Delta/\Delta}$ LCs display an increased turnover (Publication II, Fig. 5B), which possibly accelerates their loss by diluting residual miRNA molecules. Thus, miRNA-deficiency in LCs causes perturbed cell cycle and apoptosis regulation.

There is substantial evidence that miRNAs regulate apoptosis (sections 4.3.3 and 4.3.4). Asirvatham et al. have analyzed which genes with roles in the immune system are heavily targeted by miRNAs, and identified 'hubs', genes with 8 or more miRNA target sites in their 3'UTR (Asirvatham et al., 2008). There are also three hubs amongst the apoptosis-regulating genes. The gene *bcl2l11*, encoding for the pro-apoptotic Bim, is one of these hubs. The other identified hubs are the pro-apoptotic gene *dedd* and the anti-apoptotic gene *bcl2*. Our microarray data show, although not statistically significant due to small replicate numbers, that the pro-apoptotic Bim and Dedd mRNAs are consistently increased in Dicer$^{\Delta/\Delta}$ compared to Dicer$^{wt/wt}$ LCs (~6- and ~5-fold, respectively); unfortunately, the data do not allow assessment of *bcl2* expression, as bcl2 mRNA levels are not consistently up- or downregulated (microarray data published in the ArrayExpress databank under accession number E-MEXP-2387). Nevertheless, lack of miRNA regulation seems to cause a disbalance of pro- and anti-apoptotic molecules in Dicer$^{\Delta/\Delta}$ LCs, leading to the induction of apoptosis.

In support of our findings, a similar observation has been made in Dicer-deficient B cells. Due to increased Bim levels the development of Dicer-deficient B cells was blocked (Koralov et al., 2008). Increased Bim levels were shown to result from lack of miRNAs of the related clusters miR-17~92, miR-106a~363 and miR-106b~25. Importantly, a role for Bim has also been shown in the regulation of apoptosis in DCs (Chen et al., 2007b), indicating that our findings are relevant to DC biology. Thus, miRNA-regulation of Bim can have a substantial influence on the survival of immune cells, and upregulation of Bim in miRNA-deficient Dicer$^{\Delta/\Delta}$ LCs very likely triggers apoptosis.

The increased turnover we observed in LCs (Publication II, Fig. 5B) is presumably due to dysregulation of genes involved in cell cycle regulation. Jurkin et al. have compared miRNA expression profiles of *in vitro* generated human LCs and GM-CSF DCs (Jurkin et al., 2010). Their data reveal that miRNAs of the miR-17~92 and related clusters miR-106a~363 and miR-106b~25 are more strongly expressed in LCs than in GM-CSF DCs. Our gene expression analysis shows that several genes that are involved in cell cycle regulation and are known targets of the above mentioned miRNAs, are upregulated in Dicer$^{\Delta/\Delta}$ LCs. We detected upregulation of Rbl2 (~3-fold), E2F1 (~3-fold) and PTEN (~4-fold) transcripts, all of which promote cell cycle progression. This implies a role for miR-17~92 and related miRNAs in the regulation of LC turnover.

The phenotype of miRNA-deficient immature LCs is characterized by increased proliferation and apoptosis (Publication II), both of which are cellular processes frequently targeted by miRNAs and, specifically, by miRNAs of the miR-17~92 and related clusters (reviewed in Xiao and Rajewsky, 2009). Together with the finding that miR-17~92 and related clusters are strongly expressed in LCs (Jurkin et al., 2010), this leads to the assumption that the three clusters (miR-17~92, miR-106a~363 and miR-106b~25) play a major role in LC homeostasis and might be responsible for the slow turnover and long lifespan of LCs.

7.2.2 miRNAs fine-tune DC maturation and function

The ability of DCs to induce adaptive immune responses is dependent on their maturation state. DC maturation comprises a variety of processes, such as increased surface expression of MHC I and II, downregulation of antigen uptake and processing, upregulation of the costimulatory molecules CD40, CD80 and CD86, surface expression of chemokine receptors, and secretion of chemokines and cytokines (reviewed in Banchereau et al., 2000). As described in section 4.1.4, activated DCs deliver three signals to optimally tune T cells responses to the specific type of stimulus. A first indication that miRNAs play a role in T cell priming was provided by Rodriguez et al., who showed that miR-155-deficient DCs fail to efficiently induce T cell proliferation (Rodriguez et al., 2007).

7.2.2.1 miRNAs regulate DC antigen presentation

We have analyzed whether a global lack of miRNAs affects the phenotype and function of LCs. In steady-state Dicer$^{\Delta/\Delta}$ LCs, we observed normal surface expression of many typical LC markers, including CD11c, CD24, MHC II, CD40 and CD80 (Publication II, Fig. 6B and data not shown). In contrast, expression of the receptors TGF-βRII and langerin was slightly reduced (Publication II, Fig. 3E). Logically, a reduction of surface TGF-βRII or langerin cannot be directly linked to loss of miRNA suppression. However, genes of the TGF-β pathway are frequent targets of miRNAs (Asirvatham et al., 2008) and thus TGF-βRII expression might be indirectly controlled by miRNAs. As a lack of TGF-β signaling in LCs is accompanied by spontaneous LC maturation (Kel et al., 2010), and we do not see this in Dicer$^{\Delta/\Delta}$ mice, we assume that reduced amounts of surface TGF-βRII do not severely affect LC function. It might be possible, however, that reduced TGF-β signaling accounts for reduced expression of langerin, as langerin expression in *in vitro* DC cultures has been shown to depend on supplementation with TGF-β (Valladeau et al., 2002). Whether the reduced level of langerin expression, which also accounts for the absence of Birbeck granules, represents a disadvantage for Dicer$^{\Delta/\Delta}$ mice in anti-viral defense (section 4.1.1.2), has not yet been analyzed.

Intriguingly, we uncovered a failure to upregulate certain surface markers in *ex vivo* activated Dicer$^{\Delta/\Delta}$ LCs. While CCR7 was properly upregulated to allow for LC migration, and also MHC I surface expression was induced normally, the maturation markers MHC II, CD40 and CD86 were not fully upregulated (Publication II, Fig. 6B). As the maturation defect was restricted to certain molecules, we can conclude that there is no global block of, for example, transcription or translation, but a defect in specific pathways that are under miRNA control. A defect in the surface transportation of these molecules, however, cannot be excluded. Importantly, this defect hindered Dicer$^{\Delta/\Delta}$ LCs to prime naïve T cells. In particular, Dicer$^{\Delta/\Delta}$ LCs were capable of inducing CD8$^+$ T cell proliferation, but they failed to induce CD4$^+$ T cell proliferation (Publication II, Fig. 6C). The fact that protein-derived peptides could be cross-presented via MHC I, but could not be efficiently presented via MHC II, indicates that antigen uptake is not compromised in Dicer$^{\Delta/\Delta}$ LCs. However, as several key signaling molecules are verified miRNA targets (reviewed in O'Neill et al., 2011), we assume that components and regulators of the various TLR and IL-1 signaling pathways, which lead to transcription factor activation and DC maturation, are dysregulated in the absence of miRNAs.

Our data show that only antigen presentation via MHC II, and not via MHC I, is affected by loss of miRNAs. MHC I and MHC II presentation pathways differ significantly (reviewed in Vyas et al., 2008). They are differentially regulated, require different enzymes for protein degradation and involve different endosomal compartments. Notably, MHC I is expressed on all nucleated cells, while MHC II expression is mostly restricted to APCs. As different cell types display different miRNA expression patterns, miRNA-regulation of MHC I-regulating genes seems unlikely. In contrast, innate immune cells, including DCs, in which a set of miRNAs such as miR-155, miR-146 and miR-21 are ubiquitously expressed (reviewed in O'Neill et al., 2011), might have adapted the 3'UTRs of genes involved in MHC II presentation to allow for miRNA regulation. Thus, miRNA regulation of the MHC II pathway may have evolved particularly in APCs to add another layer of control in order to optimize adaptive immune responses.

miRNA control of MHC II antigen presentation is conceivable due to the complexity of this process. Many steps are involved, such as protein degradation, peptide loading onto MHC II complexes and the delivery of peptide-MHC II complexes to the surface (reviewed in van Niel et al., 2008). The gene encoding MHC II itself does not contain miRNA target sites, but target sites were found in several genes that regulate MHC II expression (Asirvatham et al., 2008). Due to technical constraints when working with LCs, we performed several experiments with Dicer$^{\Delta/\Delta}$ BM-derived DCs. In this way we found reduced mRNA levels of the MHC II master regulator CIITA and similarly reduced MHC II mRNA levels in Dicer$^{\Delta/\Delta}$ DCs compared to Dicer$^{wt/wt}$ DCs (Table 1). Therefore, reduced transcription of MHC II might at least partially account for reduced surface MHC II levels in Dicer$^{\Delta/\Delta}$ LCs. Reduced CIITA levels may be explained by derepression of a CIITA inhibitor, such as Smad3 (Dong et al., 2001), which indeed seems to contain miRNA binding sites (Asirvatham et al., 2008). Another plausible mechanism involves the ubiquitin ligase MARCH1. Upon DC maturation, MHC II is stabilized at the surface, which is the result of downregulation of MARCH1 (De Gassart et al., 2008). We have speculated that loss of miRNAs in Dicer$^{\Delta/\Delta}$ LCs leads to unblocking of MARCH1, which TargetScan predicts to have a binding site for miR-9. Supporting this notion, MARCH1 is strongly upregulated in LPS-stimulated Dicer$^{\Delta/\Delta}$ DCs (Table 1). miR-9 has been shown to be upregulated upon maturation of macrophages and neutrophils (Bazzoni et al., 2009). However, we could not confirm upregulation of miR-9 in mature DCs (Table 1), but a miRNA-independent posttranscriptional regulation of MARCH1 has been suggested (Jabbour et al.,

2009) and might play a role here. Nevertheless, miR-9 might function in myeloid cells other than DCs to stabilize surface MHC II by downregulating MARCH1. Apart from the failure in surface upregulation of MHC II molecules, we also observed a failure in the upregulation of CD40 and CD86 in Dicer$^{\Delta/\Delta}$ LCs, suggesting that miRNA-deficiency generally affects the LC maturation process by interfering with multiple signaling pathways.

7.2.2.2 miR-146a and miR-155 are key miRNAs in DC maturation

Two particular miRNAs have emerged as key modulators of DC maturation: miR-146a and miR-155. Both are highly upregulated upon TLR signaling in a NF-κB-dependent manner (reviewed in O'Neill et al., 2011). In view of very recent findings (Dunand-Sauthier et al., 2011; Lu et al., 2011), the maturation defect of Dicer$^{\Delta/\Delta}$ LCs appears to be at least partially due to the absence of miR-155. Initially, Rodriguez et al. demonstrated indirectly that miR-155 plays a role for DC function, as miR-155-deficient DCs failed to efficiently active T cells (Rodriguez et al., 2007). This finding was confirmed by another group (Dunand-Sauthier et al., 2011). However, regarding the question as to whether miR-155 is required for upregulation of MHC II and costimulatory molecules, there have been contradictory findings. Rodriguez et al. found 'similar' levels of MHC II and CD86 in activated miR-155-deficient DCs. However, in their data, miR-155-deficient DCs appear to have lower levels of MHC II and CD86, although these differences may not be statistically significant (Rodriguez et al., 2007). Data from Dunand-Sauthier et al. support the notion that miR-155 is important for upregulation of the markers MHC II, CD40 and CD86 (Dunand-Sauthier et al., 2011). In striking contrast, Lu et al. found that miR-155-deficiency does not account for reduced expression of MHC II or costimulatory molecules (Lu et al., 2011). The three different groups drew their conclusions from comparable experiments with LPS-stimulated BM-derived GM-CSF DCs. We have performed similar experiments with Dicer$^{\Delta/\Delta}$ BM-derived DCs, which mostly indicated reduced upregulation of MHC II, CD40 and CD86 compared to Dicer$^{wt/wt}$ DCs (unpublished observations), in line with results from Rodriguez et al. and Dunand-Sauthier et al.

miR-155 is believed to enhance pro-inflammatory responses as there are several negative regulators of inflammatory responses amongst the known targets of miR-155 (reviewed in O'Neill et al., 2011). For example the phosphatase SHIP1, which negatively regulates TLR signaling (An et al., 2005), has been shown to be a target of miR-155 (O'Connell et al., 2009). Another miR-155 target that is likely to be involved in DC maturation was identified by Lu et al., who showed that miR-155-deficiency leads to derepression of SOCS1, an inhibitor of Janus kinase/signal transducer and activator of transcription (JAK/STAT) signaling, and probably thereby causes reduced IL-12p70 cytokine production in mature DCs (Lu et al., 2011). miR-155-mediated silencing of c-fos is another verified mechanism which, when abrogated, impairs DC maturation and IL-12 production (Dunand-Sauthier et al., 2011). Thus, it may be that Dicer$^{\Delta/\Delta}$ LCs, which cannot upregulate miR-155, are inefficient in activating T cells because they do not provide sufficient amounts of IL-12. IL-12, which can deliver a 'signal 3' to T cells, plays an important role in T_H1 differentiation and CTL function (section 4.1.4). However, it cannot be excluded that miR-155, besides its pro-inflammatory function, also has anti-inflammatory effects, as miR-155 targeting of TAB2, a component of the TLR/IL-1 signaling pathway, has been observed (Ceppi et al., 2009). Although the impact of miR-155 on MHC II and costimulatory molecules is not yet completely defined, overall the existing data suggest that upregulation of miR-155 upon DC activation is necessary to silence genes that would otherwise interfere with signaling pathways which lead to DC maturation and IL-12 production. Thus, miR-155-deficiency, as well as Dicer-deficiency, appears to inhibit complete DC maturation and thus the DCs' T cell priming function.

To date there is no description of DCs lacking miR-146. However, it is known that NF-κB induces miR-146a in response to TLR signaling in macrophages (section 4.3.3). In turn, miR-146a functions in a negative feedback loop and silences the TLR signaling components IRAK1, IRAK2 and TRAF6 to prevent excessive inflammation. Interestingly, LCs constitutively express higher levels of miR-146a than other DC subsets, which corresponds with lower NF-κB activation (Jurkin et al., 2010). Inversely, silencing of miR-146a in LCs has been shown to cause enhanced NF-κB activation. NF-κB levels, however, are not only regulated indirectly by for example miR-146a, but also directly, as NF-κB1 transcripts contain a binding site for miR-9 (Bazzoni et al., 2009), such that higher NF-κB1 levels would be expected in Dicer$^{\Delta/\Delta}$ LCs. Indeed, in LPS-stimulated Dicer$^{\Delta/\Delta}$ DCs we have observed higher NF-κB1 transcript levels than in Dicer$^{wt/wt}$ DCs (Table 1). Nevertheless, whether a lack of miR-146a and miR-9 actually influences NF-κB activation is still unclear. We speculate that it is predominantly the lack of miR-155, which is the miRNA most strongly upregulated upon DC stimulation (Lu et al., 2011), that determines the phenotype of Dicer$^{\Delta/\Delta}$ LCs, by increasing levels of inhibitory molecules. Other possible effects of global miRNA-deficiency, for example higher NF-κB activation caused by lack of miR-146a and miR-9, may be masked or suppressed by the dominant action of genes having a miR-155 signature. Gene expression profiling performed with stimulated LCs and subsequent analysis of miRNA signatures (as performed in

(Koralov et al., 2008)) would possibly reveal the miRNAs, whose absence contributes most to the maturation defect.

Table 1: Expression profile of selected mRNAs and miRNAs in Dicer$^{wt/wt}$ and Dicer$^{\Delta/\Delta}$ BM-derived DCs during LPS-induced activation

		t0	t2	t20
MHC I	Dicer$^{wt/wt}$	100%	177%	315%
	Dicer$^{\Delta/\Delta}$	76%	120%	370%
MHC II	Dicer$^{wt/wt}$	100%	158%	63%
	Dicer$^{\Delta/\Delta}$	51%	90%	21%
CIITA	Dicer$^{wt/wt}$	100%	184%	135%
	Dicer$^{\Delta/\Delta}$	43%	83%	42%
MARCH1	Dicer$^{wt/wt}$	100%	96%	114%
	Dicer$^{\Delta/\Delta}$	180%	138%	711%
NF-κB1	Dicer$^{wt/wt}$	100%	284%	150%
	Dicer$^{\Delta/\Delta}$	83%	472%	220%
miR-9	Dicer$^{wt/wt}$	100%	88%	54%
	Dicer$^{\Delta/\Delta}$	13%	9%	9%
miR-155	Dicer$^{wt/wt}$	100%	109%	306%
	Dicer$^{\Delta/\Delta}$	13%	20%	49%

FACS sorted CD11c$^+$ BM-derived DCs either non-stimulated (t0), or stimulated with 100 ng/ml LPS for 2 h (t2) or 20 h (t20) were analyzed for indicated mRNA and miRNA expression levels by RT-PCR as described in Publication I. Values indicate the mean expression of two independent samples and are expressed as percentage of non-stimulated Dicer$^{wt/wt}$ levels.

Asirvatham et al. have reported that proteins involved in signal transduction are the primary targets of miRNAs (Asirvatham et al., 2008). Therefore it can be assumed that global miRNA-deficiency causes a considerable perturbation of signaling pathways, which we have proven in Dicer$^{\Delta/\Delta}$ LCs and GM-CSF DCs. We have demonstrated a failure of DCs to properly integrate environmental signals in the absence of miRNAs, which becomes evident in the incomplete upregulation of surface MHC II and costimulatory molecules. Thus, miRNAs are crucial for fine-tuning the DC maturation process and consequently control the induction of appropriate T cell responses.

7.3 Outlook – Implications for immunotherapy

DCs are powerful regulators of the immune system and therefore are considered as attractive tools for immunotherapy of cancer or autoimmune diseases (section 4.1.5). However, DC-based immunotherapy has only achieved limited success so far (Huang et al., 2011), making new approaches desirable. In cancer treatment, for example, DCs with enhanced immunogenicity would be required to overcome an immunosuppressive tumor environment. Therefore, strategies to selectively manipulate DCs towards an optimized function are under consideration. The idea not to only modulate the expression level of a single protein, but to manipulate multiple proteins in order to enhance DC efficacy, is appealing. miRNAs have become a research focus as potential therapeutic agents, as a single miRNA can target a large number of genes or gene regulatory networks that are often dysregulated in cancer, such as apoptosis, cell cycle or genomic stability (reviewed in Garzon et al., 2010).

Our studies were the first to show that miRNAs can critically influence phenotypic DC maturation (Publication II). Moreover, we have shown that miRNAs can alter DC functionality, which has been demonstrated by others as well (Ceppi et al., 2009; Dunand-Sauthier et al., 2011; Jurkin et al., 2010; Lu et al., 2011; Rodriguez et al., 2007). These existing data suggest that it may be possible to boost DC immunogenicity by increasing DC levels of pro-inflammatory miRNAs and/or decreasing the levels of anti-inflammatory miRNAs. This strategy to enhance DC immunogenicity could be exploited, for example, in DC-based tumor vaccines using *ex vivo* antigen-loaded DCs (section 4.1.5.1). Furthermore, the identification and modulation of miRNAs that help to tune the DC cytokine profiles towards a T_H1 or T_H2 profile could lead to more successful treatment of, for example, cancer or allergic diseases (reviewed in Steinman and Banchereau, 2007).

miR-155 represents an interesting miRNA for DC manipulation, as it plays a major role in fine-tuning the DC maturation process (reviewed in O'Neill et al., 2011). Ectopic expression of miR-155 in DCs used as vaccines might prevent the DCs from becoming tolerogenic in an immunosuppressive tumor environment. IL-10 is one of the tumor-secreted factors that has a crucial role in inhibiting successful DC-based immunotherapy (reviewed in Huang et al., 2011). Notably, IL-10 has been shown to inhibit miR-155 induction (McCoy et al., 2010) and is known to render DCs tolerogenic by inducing their expression of anti-inflammatory SOCS molecules. By targeting SOCS1 (Lu et al., 2011), artificially boosted miR-155 action in DCs might put a break on the negative feedback loop resulting from IL-10 signaling, and enhance expression of MHC II and costimulatory molecules, as well as IL-12 production (section 7.2.2.2).

However, translation of miRNA *in vitro* studies to *in vivo* applications is difficult and appears even more complicated in light of a recent report by Mao et al. (Mao et al., 2011). They showed that ectopic miR-155 expression in *in vivo* LCs inhibits T cell responses, which is contradictory to previous *in vitro* findings indicating a pro-inflammatory role for miR-155. Their results need careful interpretation however, as they used a biolistic approach which can target other cells in addition to LCs.

The hitherto existing research on miRNA involvement in DC maturation indicates that specifying the function of individual miRNAs *in vitro* and *in vivo* would be worthwhile to gain a better understanding of the contribution of individual miRNAs to DC maturation. Based on these prospective results it might be possible to exploit the function of particular miRNAs that, for example, silence signaling pathways interfering with DC maturation, in order to boost current DC-based therapeutics.

8 References

Aarntzen, E.H.J.G., Figdor, C.G., Adema, G.J., Punt, C.J.A., and Vries, I.J.M. (2008). Dendritic cell vaccination and immune monitoring. *Canc Immunol Immunother* 57, 1559-1568.

Akashi, K., Traver, D., Miyamoto, T., and Weissman, I.L. (2000). A clonogenic common myeloid progenitor that gives rise to all myeloid lineages. *Nature* 404, 193-197.

Akira, S., Uematsu, S., and Takeuchi, O. (2006). Pathogen Recognition and Innate Immunity. *Cell* 124, 783-801.

Ambros, V. (2008). The evolution of our thinking about microRNAs. *Nat Med* 14, 1036-1040.

An, H., Xu, H., Zhang, M., Zhou, J., Feng, T., Qian, C., Qi, R., and Cao, X. (2005). Src homology 2 domain-containing inositol-5-phosphatase 1 (SHIP1) negatively regulates TLR4-mediated LPS response primarily through a phosphatase activity- and PI-3K-independent mechanism. *Blood* 105, 4685-4692.

Asirvatham, A.J., Gregorie, C.J., Hu, Z., Magner, W.J., and Tomasi, T.B. (2008). MicroRNA targets in immune genes and the Dicer/Argonaute and ARE machinery components. *Mol Immunol* 45, 1995-2006.

Baek, D., Villén, J., Shin, C., Camargo, F.D., Gygi, S.P., and Bartel, D.P. (2008). The impact of microRNAs on protein output. *Nature* 455, 64-71.

Baltimore, D., Boldin, M.P., O'Connell, R.M., Rao, D.S., and Taganov, K.D. (2008). MicroRNAs: new regulators of immune cell development and function. *Nat Immunol* 9, 839-845.

Banchereau, J., Briere, F., Caux, C., Davoust, J., Lebecque, S., Liu, Y.J., Pulendran, B., and Palucka, K. (2000). Immunobiology of dendritic cells. *Annu Rev Immunol* 18, 767-811.

Bartel, D.P. (2009). MicroRNAs: Target Recognition and Regulatory Functions. *Cell* 136, 215-233.

Bazzoni, F., Rossato, M., Fabbri, M., Gaudiosi, D., Mirolo, M., Mori, L., Tamassia, N., Mantovani, A., Cassatella, M.A., and Locati, M. (2009). Induction and regulatory function of miR-9 in human monocytes and neutrophils exposed to proinflammatory signals. *Proc Natl Acad Sci U S A* 106, 5282-5287.

Bedoui, S., Whitney, P.G., Waithman, J., Eidsmo, L., Wakim, L., Caminschi, I., Allan, R.S., Wojtasiak, M., Shortman, K., Carbone, F.R., *et al.* (2009). Cross-presentation of viral and self antigens by skin-derived CD103+ dendritic cells. *Nat Immunol* 10, 488-495.

Bernstein, E., Kim, S.Y., Carmell, M.A., Murchison, E.P., Alcorn, H., Li, M.Z., Mills, A.A., Elledge, S.J., Anderson, K.V., and Hannon, G.J. (2003). Dicer is essential for mouse development. *Nat Genet* 35, 215-217.

Birnberg, T., Baron, L., Sapoznikov, A., Caton, M., Cervantesbarragan, L., Makia, D., Krauthgamer, R., Brenner, O., Ludewig, B., and Brockschnieder, D. (2008). Lack of Conventional Dendritic Cells Is Compatible with Normal Development and T Cell Homeostasis, but Causes Myeloid Proliferative Syndrome. *Immunity* 29, 986-997.

Bonehill, A., Van Nuffel, A.M., Corthals, J., Tuyaerts, S., Heirman, C., Francois, V., Colau, D., van der Bruggen, P., Neyns, B., and Thielemans, K. (2009). Single-step antigen loading and activation of dendritic cells by mRNA electroporation for the purpose of therapeutic vaccination in melanoma patients. *Clin Cancer Res* 15, 3366-3375.

Bozzacco, L., Trumpfheller, C., Siegal, F.P., Mehandru, S., Markowitz, M., Carrington, M., Nussenzweig, M.C., Piperno, A.G., and Steinman, R.M. (2007). DEC-205 receptor on dendritic cells mediates presentation of HIV gag protein to CD8+ T cells in a spectrum of human MHC I haplotypes. *Proc Natl Acad Sci U S A* 104, 1289-1294.

Bursch, L.S., Wang, L., Igyarto, B., Kissenpfennig, A., Malissen, B., Kaplan, D.H., and Hogquist, K.A. (2007). Identification of a novel population of Langerin+ dendritic cells. *J Exp Med* 204, 3147-3156.

Calin, G.A., Dumitru, C.D., Shimizu, M., Bichi, R., Zupo, S., Noch, E., Aldler, H., Rattan, S., Keating, M., Rai, K., et al. (2002). Frequent deletions and down-regulation of micro- RNA genes miR15 and miR16 at 13q14 in chronic lymphocytic leukemia. *Proc Natl Acad Sci U S A* 99, 15524-15529.

Calin, G.A., Sevignani, C., Dumitru, C.D., Hyslop, T., Noch, E., Yendamuri, S., Shimizu, M., Rattan, S., Bullrich, F., Negrini, M., et al. (2004). Human microRNA genes are frequently located at fragile sites and genomic regions involved in cancers. *Proc Natl Acad Sci U S A* 101, 2999-3004.

Caminschi, I., Lahoud, M.H., and Shortman, K. (2009). Enhancing immune responses by targeting antigen to DC. *Eur J Immunol* 39, 931-938.

Caminschi, I., Proietto, A.I., Ahmet, F., Kitsoulis, S., Shin Teh, J., Lo, J.C.Y., Rizzitelli, A., Wu, L., Vremec, D., van Dommelen, S.L.H., et al. (2008). The dendritic cell subtype-restricted C-type lectin Clec9A is a target for vaccine enhancement. *Blood* 112, 3264-3273.

Carotta, S., Dakic, A., D'Amico, A., Pang, S.H.M., Greig, K.T., Nutt, S.L., and Wu, L. (2010). The Transcription Factor PU.1 Controls Dendritic Cell Development and Flt3 Cytokine Receptor Expression in a Dose-Dependent Manner. *Immunity* 32, 628-641.

Ceppi, M., Pereira, P.M., Dunand-Sauthier, I., Barras, E., Reith, W., Santos, M.A., and Pierre, P. (2009). MicroRNA-155 modulates the interleukin-1 signaling pathway in activated human monocyte-derived dendritic cells. *Proc Natl Acad Sci U S A* 106, 2735-2740.

Chatterjee, S., and Grosshans, H. (2009). Active turnover modulates mature microRNA activity in Caenorhabditis elegans. *Nature* 461, 546-549.

Chen, C.Z. (2004). MicroRNAs Modulate Hematopoietic Lineage Differentiation. *Science* 303, 83-86.

Chen, M. (2006). Dendritic Cell Apoptosis in the Maintenance of Immune Tolerance. *Science* 311, 1160-1164.

Chen, M., Huang, L., Shabier, Z., and Wang, J. (2007a). Regulation of the lifespan in dendritic cell subsets. *Mol Immunol* 44, 2558-2565.

Chen, M., Huang, L., and Wang, J. (2007b). Deficiency of Bim in dendritic cells contributes to overactivation of lymphocytes and autoimmunity. *Blood* 109, 4360-4367.

Cheng, P., Zhou, J., and Gabrilovich, D. (2010). Regulation of dendritic cell differentiation and function by Notch and Wnt pathways. *Immunol Rev* 234, 105-119.

Chorro, L., Sarde, A., Li, M., Woollard, K.J., Chambon, P., Malissen, B., Kissenpfennig, A., Barbaroux, J.B., Groves, R., and Geissmann, F. (2009). Langerhans cell (LC) proliferation mediates neonatal development, homeostasis, and inflammation-associated expansion of the epidermal LC network. *J Exp Med* 206, 3089-3100.

Cisse, B., Caton, M., Lehner, M., Maeda, T., Scheu, S., Locksley, R., Holmberg, D., Zweier, C., Denhollander, N., and Kant, S. (2008). Transcription Factor E2-2 Is an Essential and Specific Regulator of Plasmacytoid Dendritic Cell Development. *Cell* 135, 37-48.

Cobb, B.S. (2005). T cell lineage choice and differentiation in the absence of the RNase III enzyme Dicer. *J Exp Med* 201, 1367-1373.

Cobb, B.S., Hertweck, A., Smith, J., O'Connor, E., Graf, D., Cook, T., Smale, S.T., Sakaguchi, S., Livesey, F.J., Fisher, A.G., et al. (2006). A role for Dicer in immune regulation. *J Exp Med* 203, 2519-2527.

Collin, M.P., Hart, D.N., Jackson, G.H., Cook, G., Cavet, J., Mackinnon, S., Middleton, P.G., and Dickinson, A.M. (2006). The fate of human Langerhans cells in hematopoietic stem cell transplantation. *J Exp Med* 203, 27-33.

Colonna, M., Trinchieri, G., and Liu, Y.-J. (2004). Plasmacytoid dendritic cells in immunity. *Nat Immunol* 5, 1219-1226.

Coquerelle, C., and Moser, M. (2010). DC subsets in positive and negative regulation of immunity. *Immunol Rev* 234, 317-334.

Costinean, S., Zanesi, N., Pekarsky, Y., Tili, E., Volinia, S., Heerema, N., and Croce, C.M. (2006). Pre-B cell proliferation and lymphoblastic leukemia/high-grade lymphoma in E(mu)-miR155 transgenic mice. *Proc Natl Acad Sci U S A* 103, 7024-7029.

Crozat, K., Guiton, R., Guilliams, M., Henri, S., Baranek, T., Schwartz-Cornil, I., Malissen, B., and Dalod, M. (2010). Comparative genomics as a tool to reveal functional equivalences between human and mouse dendritic cell subsets. *Immunol Rev* 234, 177-198.

Dai, X.M., Ryan, G.R., Hapel, A.J., Dominguez, M.G., Russell, R.G., Kapp, S., Sylvestre, V., and Stanley, E.R. (2002). Targeted disruption of the mouse colony-stimulating factor 1 receptor gene results in osteopetrosis, mononuclear phagocyte deficiency, increased primitive progenitor cell frequencies, and reproductive defects. *Blood* 99, 111-120.

Davis, B.N., and Hata, A. (2009). Regulation of MicroRNA Biogenesis: A miRiad of mechanisms. *Cell Commun Signal* 7, 18.

De Gassart, A., Camossetto, V., Thibodeau, J., Ceppi, M., Catalan, N., Pierre, P., and Gatti, E. (2008). MHC class II stabilization at the surface of human dendritic cells is the result of maturation-dependent MARCH I down-regulation. *Proc Natl Acad Sci U S A* 105, 3491-3496.

de Witte, L., Nabatov, A., Pion, M., Fluitsma, D., de Jong, M.A., de Gruijl, T., Piguet, V., van Kooyk, Y., and Geijtenbeek, T.B. (2007). Langerin is a natural barrier to HIV-1 transmission by Langerhans cells. *Nat Med* 13, 367-371.

del Rio, M.L., Rodriguez-Barbosa, J.I., Kremmer, E., and Forster, R. (2007). CD103- and CD103+ bronchial lymph node dendritic cells are specialized in presenting and cross-presenting innocuous antigen to CD4+ and CD8+ T cells. *J Immunol* 178, 6861-6866.

den Haan, J.M., Lehar, S.M., and Bevan, M.J. (2000). CD8(+) but not CD8(-) dendritic cells cross-prime cytotoxic T cells in vivo. *J Exp Med* 192, 1685-1696.

Dominguez, P.M., and Ardavin, C. (2010). Differentiation and function of mouse monocyte-derived dendritic cells in steady state and inflammation. *Immunol Rev* 234, 90-104.

Dong, Y., Tang, L., Letterio, J.J., and Benveniste, E.N. (2001). The Smad3 protein is involved in TGF-beta inhibition of class II transactivator and class II MHC expression. *J Immunol* 167, 311-319.

Douillard, P., Stoitzner, P., Tripp, C.H., Clair-Moninot, V., Ait-Yahia, S., McLellan, A.D., Eggert, A., Romani, N., and Saeland, S. (2005). Mouse lymphoid tissue contains distinct subsets of langerin/CD207 dendritic cells, only one of which represents epidermal-derived Langerhans cells. *J Investig Dermatol* 125, 983-994.

Dunand-Sauthier, I., Santiago-Raber, M.L., Capponi, L., Vejnar, C.E., Schaad, O., Irla, M., Seguin-Estevez, Q., Descombes, P., Zdobnov, E.M., Acha-Orbea, H., *et al.* (2011). Silencing of c-Fos expression by microRNA-155 is critical for dendritic cell maturation and function. *Blood* 117, 4490-4500.

Edelson, B.T., Kc, W., Juang, R., Kohyama, M., Benoit, L.A., Klekotka, P.A., Moon, C., Albring, J.C., Ise, W., Michael, D.G., *et al.* (2010). Peripheral CD103+ dendritic cells form a unified subset developmentally related to CD8 + conventional dendritic cells. *J Exp Med* 207, 823-836.

Fancke, B., Suter, M., Hochrein, H., and O'Keeffe, M. (2008). M-CSF: a novel plasmacytoid and conventional dendritic cell poietin. *Blood* 111, 150-159.

Fedeli, M., Napolitano, A., Wong, M.P.M., Marcais, A., de Lalla, C., Colucci, F., Merkenschlager, M., Dellabona, P., and Casorati, G. (2009). Dicer-Dependent MicroRNA Pathway Controls Invariant NKT Cell Development. *J Immunol* 183, 2506-2512.

Feinberg, H., Taylor, M.E., Razi, N., McBride, R., Knirel, Y.A., Graham, S.A., Drickamer, K., and Weis, W.I. (2011). Structural basis for langerin recognition of diverse pathogen and mammalian glycans through a single binding site. *J Mol Biol* 405, 1027-1039.

Felker, P., Sere, K., Lin, Q., Becker, C., Hristov, M., Hieronymus, T., and Zenke, M. (2010). TGF-β1 Accelerates Dendritic Cell Differentiation from Common Dendritic Cell Progenitors and Directs Subset Specification toward Conventional Dendritic Cells. *J Immunol* 185, 5326-5335.

Felli, N. (2005). MicroRNAs 221 and 222 inhibit normal erythropoiesis and erythroleukemic cell growth via kit receptor down-modulation. *Proc Natl Acad Sci U S A* 102, 18081-18086.

Filipowicz, W., Bhattacharyya, S.N., and Sonenberg, N. (2008). Mechanisms of post-transcriptional regulation by microRNAs: are the answers in sight? *Nat Rev Genet* 2008, 102-114.

Flacher, V., Douillard, P., Ait-Yahia, S., Stoitzner, P., Clair-Moninot, V., Romani, N., and Saeland, S. (2008). Expression of langerin/CD207 reveals dendritic cell heterogeneity between inbred mouse strains. *Immunology* 123, 339-347.

Forster, I., and Rajewsky, K. (1990). The bulk of the peripheral B-cell pool in mice is stable and not rapidly renewed from the bone marrow. *Proc Natl Acad Sci U S A* 87, 4781-4784.

Friedman, R.C., Farh, K.K.H., Burge, C.B., and Bartel, D.P. (2008). Most mammalian mRNAs are conserved targets of microRNAs. *Genome Res* 19, 92-105.

Fujita, J., Mizuki, M., Otsuka, M., Ezoe, S., Tanaka, H., Satoh, Y., Fukushima, K., Tokunaga, M., Matsumura, I., and Kanakura, Y. (2011). Myeloid neoplasm-related gene abnormalities differentially affect dendritic cell differentiation from murine hematopoietic stem/progenitor cells. *Immunol Lett* 136, 61-73.

Fukao, T., Fukuda, Y., Kiga, K., Sharif, J., Hino, K., Enomoto, Y., Kawamura, A., Nakamura, K., Takeuchi, T., and Tanabe, M. (2007). An evolutionarily conserved mechanism for microRNA-223 expression revealed by microRNA gene profiling. *Cell* 129, 617-631.

Gantier, M.P., McCoy, C.E., Rusinova, I., Saulep, D., Wang, D., Xu, D., Irving, A.T., Behlke, M.A., Hertzog, P.J., Mackay, F., et al. (2011). Analysis of microRNA turnover in mammalian cells following Dicer1 ablation. *Nucleic Acids Res* 39, 5692–5703.

Garzon, R., Garofalo, M., Martelli, M.P., Briesewitz, R., Wang, L., Fernandez-Cymering, C., Volinia, S., Liu, C.G., Schnittger, S., Haferlach, T., et al. (2008). Distinctive microRNA signature of acute myeloid leukemia bearing cytoplasmic mutated nucleophosmin. *Proc Natl Acad Sci U S A* 105, 3945-3950.

Garzon, R., Marcucci, G., and Croce, C.M. (2010). Targeting microRNAs in cancer: rationale, strategies and challenges. *Nat Rev Drug Discov* 9, 775-789.

Ghosh, H.S., Cisse, B., Bunin, A., Lewis, K.L., and Reizis, B. (2010). Continuous Expression of the Transcription Factor E2-2 Maintains the Cell Fate of Mature Plasmacytoid Dendritic Cells. *Immunity* 33, 905-916.

Gilliet, M., Boonstra, A., Paturel, C., Antonenko, S., Xu, X.L., Trinchieri, G., O'Garra, A., and Liu, Y.J. (2002). The development of murine plasmacytoid dendritic cell precursors is differentially regulated by FLT3-ligand and granulocyte/macrophage colony-stimulating factor. *J Exp Med* 195, 953-958.

Ginhoux, F., Collin, M.P., Bogunovic, M., Abel, M., Leboeuf, M., Helft, J., Ochando, J., Kissenpfennig, A., Malissen, B., Grisotto, M., et al. (2007). Blood-derived dermal langerin+ dendritic cells survey the skin in the steady state. *J Exp Med* 204, 3133-3146.

Ginhoux, F., Tacke, F., Angeli, V., Bogunovic, M., Loubeau, M., Dai, X.-M., Stanley, E.R., Randolph, G.J., and Merad, M. (2006). Langerhans cells arise from monocytes in vivo. *Nat Immunol* 7, 265-273.

Guilliams, M., Henri, S., Tamoutounour, S., Ardouin, L., Schwartz-Cornil, I., Dalod, M., and Malissen, B. (2010). From skin dendritic cells to a simplified classification of human and mouse dendritic cell subsets. *Eur J Immunol* 40, 2089-2094.

Guo, H., Ingolia, N.T., Weissman, J.S., and Bartel, D.P. (2010). Mammalian microRNAs predominantly act to decrease target mRNA levels. *Nature* 466, 835-840.

Hacker, C., Kirsch, R.D., Ju, X.S., Hieronymus, T., Gust, T.C., Kuhl, C., Jorgas, T., Kurz, S.M., Rose-John, S., Yokota, Y., *et al.* (2003). Transcriptional profiling identifies Id2 function in dendritic cell development. *Nat Immunol* 4, 380-386.

Hashimi, S.T., Fulcher, J.A., Chang, M.H., Gov, L., Wang, S., and Lee, B. (2009). MicroRNA profiling identifies miR-34a and miR-21 and their target genes JAG1 and WNT1 in the coordinate regulation of dendritic cell differentiation. *Blood* 114, 404-414.

Heath, W.R., and Carbone, F.R. (2009). Dendritic cell subsets in primary and secondary T cell responses at body surfaces. *Nat Immunol* 10, 1237-1244.

Henri, S., Poulin, L.F., Tamoutounour, S., Ardouin, L., Guilliams, M., de Bovis, B., Devilard, E., Viret, C., Azukizawa, H., Kissenpfennig, A., *et al.* (2009). CD207+ CD103+ dermal dendritic cells cross-present keratinocyte-derived antigens irrespective of the presence of Langerhans cells. *J Exp Med* 207, 189-206.

Higano, C.S., Schellhammer, P.F., Small, E.J., Burch, P.A., Nemunaitis, J., Yuh, L., Provost, N., and Frohlich, M.W. (2009). Integrated data from 2 randomized, double-blind, placebo-controlled, phase 3 trials of active cellular immunotherapy with sipuleucel-T in advanced prostate cancer. *Cancer* 115, 3670-3679.

Hildner, K., Edelson, B.T., Purtha, W.E., Diamond, M., Matsushita, H., Kohyama, M., Calderon, B., Schraml, B.U., Unanue, E.R., Diamond, M.S., *et al.* (2008). Batf3 deficiency reveals a critical role for CD8alpha+ dendritic cells in cytotoxic T cell immunity. *Science* 322, 1097-1100.

Holmstrøm, K., Pedersen, A.W., Claesson, M.H., Zocca, M.-B., and Jensen, S.S. (2010). Identification of a microRNA signature in dendritic cell vaccines for cancer immunotherapy. *Hum Immunol* 71, 67-73.

Huang, F.-P., Chen, Y.-X., and To, C.K.W. (2011). Guiding the "misguided" - functional conditioning of dendritic cells for the DC-based immunotherapy against tumours. *Eur J Immunol* 41, 18-25.

Hume, D.A. (2008). Macrophages as APC and the dendritic cell myth. *J Immunol* 181, 5829-5835.

Idoyaga, J., Lubkin, A., Fiorese, C., Lahoud, M.H., Caminschi, I., Huang, Y., Rodriguez, A., Clausen, B.E., Park, C.G., Trumpfheller, C., *et al.* (2011). Comparable T helper 1 (Th1) and CD8 T-cell immunity by targeting HIV gag p24 to CD8 dendritic cells within antibodies to Langerin, DEC205, and Clec9A. *Proc Natl Acad Sci U S A* 108, 2384-2389.

Igyarto, B.Z., Jenison, M.C., Dudda, J.C., Roers, A., Muller, W., Koni, P.A., Campbell, D.J., Shlomchik, M.J., and Kaplan, D.H. (2009). Langerhans Cells Suppress Contact Hypersensitivity Responses Via Cognate CD4 Interaction and Langerhans Cell-Derived IL-10. *J Immunol* 183, 5085-5093.

Inaba, K., Inaba, M., Romani, N., Aya, H., Deguchi, M., Ikehara, S., Muramatsu, S., and Steinman, R.M. (1992). Generation of large numbers of dendritic cells from mouse bone marrow cultures supplemented with granulocyte/macrophage colony-stimulating factor. *J Exp Med* 176, 1693-1702.

Ivey, K.N., Muth, A., Arnold, J., King, F.W., Yeh, R.F., Fish, J.E., Hsiao, E.C., Schwartz, R.J., Conklin, B.R., Bernstein, H.S., *et al.* (2008). MicroRNA regulation of cell lineages in mouse and human embryonic stem cells. *Cell Stem Cell* 2, 219-229.

Iwasaki, A. (2007). Mucosal dendritic cells. *Annu Rev Immunol* 25, 381-418.

Jabbour, M., Campbell, E.M., Fares, H., and Lybarger, L. (2009). Discrete Domains of MARCH1 Mediate Its Localization, Functional Interactions, and Posttranscriptional Control of Expression. *J Immunol* 183, 6500-6512.

Johnnidis, J.B., Harris, M.H., Wheeler, R.T., Stehling-Sun, S., Lam, M.H., Kirak, O., Brummelkamp, T.R., Fleming, M.D., and Camargo, F.D. (2008). Regulation of progenitor cell proliferation and granulocyte function by microRNA-223. *Nature* 451, 1125-1129.

Jongbloed, S.L., Kassianos, A.J., McDonald, K.J., Clark, G.J., Ju, X., Angel, C.E., Chen, C.J., Dunbar, P.R., Wadley, R.B., Jeet, V., et al. (2010). Human CD141+ (BDCA-3)+ dendritic cells (DCs) represent a unique myeloid DC subset that cross-presents necrotic cell antigens. *J Exp Med* 207, 1247-1260.

Jurkin, J., Schichl, Y.M., Koeffel, R., Bauer, T., Richter, S., Konradi, S., Gesslbauer, B., and Strobl, H. (2010). miR-146a Is Differentially Expressed by Myeloid Dendritic Cell Subsets and Desensitizes Cells to TLR2-Dependent Activation. *J Immunol* 184, 4955-4965.

Kai, Z.S., and Pasquinelli, A.E. (2010). MicroRNA assassins: factors that regulate the disappearance of miRNAs. *Nat Struct Biol* 17, 5-10.

Kaplan, D.H. (2010). In vivo function of Langerhans cells and dermal dendritic cells. *Trends Immunol* 31, 446-451.

Kaplan, D.H., Kissenpfennig, A., and Clausen, B.E. (2008). Insights into Langerhans cell function from Langerhans cell ablation models. *Eur J Immunol* 38, 2369-2376.

Kaplan, D.H., Li, M.O., Jenison, M.C., Shlomchik, W.D., Flavell, R.A., and Shlomchik, M.J. (2007). Autocrine/paracrine TGF 1 is required for the development of epidermal Langerhans cells. *J Exp Med* 204, 2545-2552.

Kassner, N., Krueger, M., Yagita, H., Dzionek, A., Hutloff, A., Kroczek, R., Scheffold, A., and Rutz, S. (2010). Cutting edge: Plasmacytoid dendritic cells induce IL-10 production in T cells via the Delta-like-4/Notch axis. *J Immunol* 184, 550-554.

Katoh, T., Sakaguchi, Y., Miyauchi, K., Suzuki, T., Kashiwabara, S., and Baba, T. (2009). Selective stabilization of mammalian microRNAs by 3' adenylation mediated by the cytoplasmic poly(A) polymerase GLD-2. *Gene Dev* 23, 433-438.

Kawai, T., and Akira, S. (2006). TLR signaling. *Cell Death Differ* 13, 816-825.

Kawai, T., and Akira, S. (2010). The role of pattern-recognition receptors in innate immunity: update on Toll-like receptors. *Nat Immunol* 11, 373-384.

Kel, J.M., Girard-Madoux, M.J.H., Reizis, B., and Clausen, B.E. (2010). TGF-β Is Required To Maintain the Pool of Immature Langerhans Cells in the Epidermis. *J Immunol* 185, 3248-3255.

Kent, D., Copley, M., Benz, C., Dykstra, B., Bowie, M., and Eaves, C. (2008). Regulation of hematopoietic stem cells by the steel factor/KIT signaling pathway. *Clin Cancer Res* 14, 1926-1930.

Kingston, D., Schmid, M.A., Onai, N., Obata-Onai, A., Baumjohann, D., and Manz, M.G. (2009). The concerted action of GM-CSF and Flt3-ligand on in vivo dendritic cell homeostasis. *Blood* 114, 835-843.

Kissenpfennig, A., Ait-Yahia, S., Clair-Moninot, V., Stossel, H., Badell, E., Bordat, Y., Pooley, J.L., Lang, T., Prina, E., Coste, I., et al. (2005). Disruption of the langerin/CD207 gene abolishes Birbeck granules without a marked loss of Langerhans cell function. *Mol Cell Biol* 25, 88-99.

Kong, K.Y., Owens, K.S., Rogers, J.H., Mullenix, J., Velu, C.S., Grimes, H.L., and Dahl, R. (2010). MIR-23A microRNA cluster inhibits B-cell development. *Exp Hematol* 38, 629-640.e621.

Koralov, S., Muljo, S., Galler, G., Krek, A., Chakraborty, C., Kanellopoulou, C., Jensen, K., Cobb, B., Merkenschlager, M., and Rajewsky, N. (2008). Dicer Ablation Affects Antibody Diversity and Cell Survival in the B Lymphocyte Lineage. *Cell* 132, 860-874.

Kosik, K.S. (2010). MicroRNAs and Cellular Phenotypy. *Cell* 143, 21-26.

Krek, A., Grün, D., Poy, M.N., Wolf, R., Rosenberg, L., Epstein, E.J., MacMenamin, P., da Piedade, I., Gunsalus, K.C., Stoffel, M., et al. (2005). Combinatorial microRNA target predictions. *Nat Genet* 37, 495-500.

Krol, J., Loedige, I., and Filipowicz, W. (2010). The widespread regulation of microRNA biogenesis, function and decay. *Nat Rev Genet* 11, 597-610.

Krug, A., French, A.R., Barchet, W., Fischer, J.A., Dzionek, A., Pingel, J.T., Orihuela, M.M., Akira, S., Yokoyama, W.M., and Colonna, M. (2004). TLR9-dependent recognition of MCMV by IPC and DC generates coordinated cytokine responses that activate antiviral NK cell function. *Immunity* 21, 107-119.

Lai, E.C., Tam, B., and Rubin, G.M. (2005). Pervasive regulation of Drosophila Notch target genes by GY-box-, Brd-box-, and K-box-class microRNAs. *Gene Dev* 19, 1067-1080.

Landgraf, P., Rusu, M., Sheridan, R., Sewer, A., Iovino, N., Aravin, A., Pfeffer, S., Rice, A., Kamphorst, A.O., and Landthaler, M. (2007). A Mammalian microRNA Expression Atlas Based on Small RNA Library Sequencing. *Cell* 129, 1401-1414.

Langerhans, P. (1868). Ueber die Nerven der menschlichen Haut. *Virchows Archiv* 44, 325-337.

Lee, R.C., Feinbaum, R.L., and Ambros, V. (1993). The C. elegans heterochronic gene lin-4 encodes small RNAs with antisense complementarity to lin-14. *Cell* 75, 843-854.

Li, Q., Chau, J., Ebert, P., Sylvester, G., Min, H., Liu, G., Braich, R., Manoharan, M., Soutschek, J., and Skare, P. (2007). miR-181a Is an Intrinsic Modulator of T Cell Sensitivity and Selection. *Cell* 129, 147-161.

Liou, L.Y., Blasius, A.L., Welch, M.J., Colonna, M., Oldstone, M.B., and Zuniga, E.I. (2008). In vivo conversion of BM plasmacytoid DC into CD11b+ conventional DC during virus infection. *Eur J Immunol* 38, 3388-3394.

Liu, K., and Nussenzweig, M.C. (2010). Origin and development of dendritic cells. *Immunol Rev* 234, 45-54.

Liu, Y.J. (2005). IPC: professional type 1 interferon-producing cells and plasmacytoid dendritic cell precursors. *Annu Rev Immunol* 23, 275-306.

Lodish, H.F., Zhou, B., Liu, G., and Chen, C.-Z. (2008). Micromanagement of the immune system by microRNAs. *Nat Rev Immunol* 8, 120-130.

Lu, C., Huang, X., Zhang, X., Roensch, K., Cao, Q., Nakayama, K.I., Blazar, B.R., Zeng, Y., and Zhou, X. (2011). MiR-221 and miR-155 regulate human dendritic cell development, apoptosis and IL-12 production through targeting of p27kip1, KPC1 and SOCS-1. *Blood* 117, 4293–4303.

Lu, J., Getz, G., Miska, E.A., Alvarez-Saavedra, E., Lamb, J., Peck, D., Sweet-Cordero, A., Ebert, B.L., Mak, R.H., Ferrando, A.A., *et al.* (2005). MicroRNA expression profiles classify human cancers. *Nature* 435, 834-838.

Lu, W., Arraes, L.C., Ferreira, W.T., and Andrieu, J.M. (2004). Therapeutic dendritic-cell vaccine for chronic HIV-1 infection. *Nat Med* 10, 1359-1365.

Luckashenak, N., Schroeder, S., Endt, K., Schmidt, D., Mahnke, K., Bachmann, M.F., Marconi, P., Deeg, C.A., and Brocker, T. (2008). Constitutive Crosspresentation of Tissue Antigens by Dendritic Cells Controls CD8+ T Cell Tolerance In Vivo. *Immunity* 28, 521-532.

Mao, C.-P., He, L., Tsai, Y.-C., Peng, S., Kang, T., Pang, X., Monie, A., Hung, C.-F., and Wu, T.C. (2011). In vivo microRNA-155 expression influences antigen-specific T cell-mediated immune responses generated by DNA vaccination. *Cell Biosci* 1, 3.

Maraskovsky, E., Brasel, K., Teepe, M., Roux, E.R., Lyman, S.D., Shortman, K., and McKenna, H.J. (1996). Dramatic increase in the numbers of functionally mature dendritic cells in Flt3 ligand-treated mice: multiple dendritic cell subpopulations identified. *J Exp Med* 184, 1953-1962.

McCoy, C.E., Sheedy, F.J., Qualls, J.E., Doyle, S.L., Quinn, S.R., Murray, P.J., and O'Neill, L.A. (2010). IL-10 inhibits miR-155 induction by toll-like receptors. *J Biol Chem* 285, 20492-20498.

Melief, C. (2008). Cancer Immunotherapy by Dendritic Cells. *Immunity* 29, 372-383.

Mendell, J.T. (2008). miRiad Roles for the miR-17-92 Cluster in Development and Disease. *Cell* 133, 217-222.

Merad, M., and Manz, M.G. (2009). Dendritic cell homeostasis. *Blood* 113, 3418-3427.

Merad, M., Manz, M.G., Karsunky, H., Wagers, A., Peters, W., Charo, I., Weissman, I.L., Cyster, J.G., and Engleman, E.G. (2002). Langerhans cells renew in the skin throughout life under steady-state conditions. *Nat Immunol* 3, 1135-1141.

Muljo, S.A. (2005). Aberrant T cell differentiation in the absence of Dicer. *J Exp Med* 202, 261-269.

Naik, S.H. (2008). Demystifying the development of dendritic cell subtypes, a little. *Immunol Cell Biol* 86, 439-452.

Naik, S.H., Proietto, A.I., Wilson, N.S., Dakic, A., Schnorrer, P., Fuchsberger, M., Lahoud, M.H., O'Keeffe, M., Shao, Q.X., Chen, W.F., *et al.* (2005). Cutting edge: generation of splenic CD8+ and CD8- dendritic cell equivalents in Fms-like tyrosine kinase 3 ligand bone marrow cultures. *J Immunol* 174, 6592-6597.

Naik, S.H., Sathe, P., Park, H.-Y., Metcalf, D., Proietto, A.I., Dakic, A., Carotta, S., O'Keeffe, M., Bahlo, M., Papenfuss, A., *et al.* (2007). Development of plasmacytoid and conventional dendritic cell subtypes from single precursor cells derived in vitro and in vivo. *Nat Immunol* 8, 1217-1226.

Neilson, J.R., Zheng, G.X.Y., Burge, C.B., and Sharp, P.A. (2007). Dynamic regulation of miRNA expression in ordered stages of cellular development. *Gene Dev* 21, 578-589.

O'Connell, R.M., Chaudhuri, A.A., Rao, D.S., and Baltimore, D. (2009). Inositol phosphatase SHIP1 is a primary target of miR-155. *Proc Natl Acad Sci U S A* 106, 7113-7118.

O'Connell, R.M., Kahn, D., Gibson, W.S.J., Round, J.L., Scholz, R.L., Chaudhuri, A.A., Kahn, M.E., Rao, D.S., and Baltimore, D. (2010a). MicroRNA-155 Promotes Autoimmune Inflammation by Enhancing Inflammatory T Cell Development. *Immunity* 33, 607-619.

O'Connell, R.M., Rao, D.S., Chaudhuri, A.A., and Baltimore, D. (2010b). Physiological and pathological roles for microRNAs in the immune system. *Nat Rev Immunol* 10, 111-122.

O'Connell, R.M., Rao, D.S., Chaudhuri, A.A., Boldin, M.P., Taganov, K.D., Nicoll, J., Paquette, R.L., and Baltimore, D. (2008). Sustained expression of microRNA-155 in hematopoietic stem cells causes a myeloproliferative disorder. *J Exp Med* 205, 585-594.

O'Neill, L.A., Sheedy, F.J., and McCoy, C.E. (2011). MicroRNAs: the fine-tuners of Toll-like receptor signalling. *Nat Rev Immunol* 11, 163-175.

O'Shea, J.J., and Paul, W.E. (2010). Mechanisms underlying lineage commitment and plasticity of helper CD4+ T cells. *Science* 327, 1098-1102.

Ohishi, K. (2001). The Notch ligand, Delta-1, inhibits the differentiation of monocytes into macrophages but permits their differentiation into dendritic cells. *Blood* 98, 1402-1407.

Ohnmacht, C., Pullner, A., King, S.B.S., Drexler, I., Meier, S., Brocker, T., and Voehringer, D. (2009). Constitutive ablation of dendritic cells breaks self-tolerance of CD4 T cells and results in spontaneous fatal autoimmunity. *J Exp Med* 206, 549-559.

Oldenhove, G., de Heusch, M., Urbain-Vansanten, G., Urbain, J., Maliszewski, C., Leo, O., and Moser, M. (2003). CD4+ CD25+ Regulatory T Cells Control T Helper Cell Type 1 Responses to Foreign Antigens Induced by Mature Dendritic Cells In Vivo. *J Exp Med* 198, 259-266.

Onai, N., Obata-Onai, A., Schmid, M.A., Ohteki, T., Jarrossay, D., and Manz, M.G. (2007). Identification of clonogenic common Flt3+M-CSFR+ plasmacytoid and conventional dendritic cell progenitors in mouse bone marrow. *Nat Immunol* 8, 1207-1216.

Palucka, A.K., Ueno, H., Connolly, J., Kerneis-Norvell, F., Blanck, J.P., Johnston, D.A., Fay, J., and Banchereau, J. (2006). Dendritic cells loaded with killed allogeneic melanoma cells can induce objective clinical responses and MART-1 specific CD8+ T-cell immunity. *J Immunother* 29, 545-557.

Perry, M.M., Williams, A.E., Tsitsiou, E., Larner-Svensson, H.M., and Lindsay, M.A. (2009). Divergent intracellular pathways regulate interleukin-1β-induced miR-146a and miR-146b expression and chemokine release in human alveolar epithelial cells. *FEBS Lett* 583, 3349-3355.

Petrocca, F., Vecchione, A., and Croce, C.M. (2008). Emerging role of miR-106b-25/miR-17-92 clusters in the control of transforming growth factor beta signaling. *Cancer Res* 68, 8191-8194.

Poliseno, L., Tuccoli, A., Mariani, L., Evangelista, M., Citti, L., Woods, K., Mercatanti, A., Hammond, S., and Rainaldi, G. (2006). MicroRNAs modulate the angiogenic properties of HUVECs. *Blood* 108, 3068-3071.

Pooley, J.L., Heath, W.R., and Shortman, K. (2001). Cutting edge: intravenous soluble antigen is presented to CD4 T cells by CD8- dendritic cells, but cross-presented to CD8 T cells by CD8+ dendritic cells. *J Immunol* 166, 5327-5330.

Poulin, L.F., Henri, S., de Bovis, B., Devilard, E., Kissenpfennig, A., and Malissen, B. (2007). The dermis contains langerin+ dendritic cells that develop and function independently of epidermal Langerhans cells. *J Exp Med* 204, 3119-3131.

Pulikkan, J.A., Dengler, V., Peramangalam, P.S., Peer Zada, A.A., Muller-Tidow, C., Bohlander, S.K., Tenen, D.G., and Behre, G. (2010). Cell-cycle regulator E2F1 and microRNA-223 comprise an autoregulatory negative feedback loop in acute myeloid leukemia. *Blood* 115, 1768-1778.

Randolph, G.J., Inaba, K., Robbiani, D.F., Steinman, R.M., and Muller, W.A. (1999). Differentiation of phagocytic monocytes into lymph node dendritic cells in vivo. *Immunity* 11, 753-761.

Randolph, G.J., Ochando, J., and Partida-Sánchez, S. (2008). Migration of Dendritic Cell Subsets and their Precursors. *Annu Rev Immunol* 26, 293-316.

Rasmussen, K.D., Simmini, S., Abreu-Goodger, C., Bartonicek, N., Di Giacomo, M., Bilbao-Cortes, D., Horos, R., Von Lindern, M., Enright, A.J., and O'Carroll, D. (2010). The miR-144/451 locus is required for erythroid homeostasis. *J Exp Med* 207, 1351-1358.

Reinhart, B.J., Slack, F.J., Basson, M., Pasquinelli, A.E., Bettinger, J.C., Rougvie, A.E., Horvitz, H.R., and Ruvkun, G. (2000). The 21-nucleotide let-7 RNA regulates developmental timing in Caenorhabditis elegans. *Nature* 403, 901-906.

Reis e Sousa, C. (2006). Dendritic cells in a mature age. *Nat Rev Immunol* 6, 476-483.

Reizis, B. (2010). Regulation of plasmacytoid dendritic cell development. *Curr Opin Immunol* 22, 206-211.

Reizis, B., Bunin, A., Ghosh, H.S., Lewis, K.L., and Sisirak, V. (2010). Plasmacytoid Dendritic Cells: Recent Progress and Open Questions. *Annu Rev Immunol* 29, 163-183.

Rettig, L., Haen, S.P., Bittermann, A.G., von Boehmer, L., Curioni, A., Kramer, S.D., Knuth, A., and Pascolo, S. (2010). Particle size and activation threshold: a new dimension of danger signaling. *Blood* 115, 4533-4541.

Robbins, S.H., Walzer, T., Dembélé, D., Thibault, C., Defays, A., Bessou, G., Xu, H., Vivier, E., Sellars, M., Pierre, P., et al. (2008). Novel insights into the relationships between dendritic cell subsets in human and mouse revealed by genome-wide expression profiling. *Genome Biol* 9, R17.

Rodriguez, A., Vigorito, E., Clare, S., Warren, M.V., Couttet, P., Soond, D.R., van Dongen, S., Grocock, R.J., Das, P.P., Miska, E.A., et al. (2007). Requirement of bic/microRNA-155 for normal immune function. *Science* 316, 608-611.

Romani, N., Clausen, B.E., and Stoitzner, P. (2010). Langerhans cells and more: langerin-expressing dendritic cell subsets in the skin. *Immunol Rev* 234, 120-141.

Sancho, D., Mourao-Sa, D., Joffre, O.P., Schulz, O., Rogers, N.C., Pennington, D.J., Carlyle, J.R., and Reis e Sousa, C. (2008). Tumor therapy in mice via antigen targeting to a novel, DC-restricted C-type lectin. *J Clin Investig* 118, 2098-2110.

Schuler, G. (2010). Dendritic cells in cancer immunotherapy. *Eur J Immunol* 40, 2123-2130.

Schuler, G., and Steinman, R.M. (1985). Murine epidermal Langerhans cells mature into potent immunostimulatory dendritic cells in vitro. *J Exp Med* 161, 526-546.

Selbach, M., Schwanhäusser, B., Thierfelder, N., Fang, Z., Khanin, R., and Rajewsky, N. (2008). Widespread changes in protein synthesis induced by microRNAs. *Nature* 455, 58-63.

Serbina, N.V., Salazar-Mather, T.P., Biron, C.A., Kuziel, W.A., and Pamer, E.G. (2003). TNF/iNOS-producing dendritic cells mediate innate immune defense against bacterial infection. *Immunity* 19, 59-70.

Sheedy, F.J., Palsson-McDermott, E., Hennessy, E.J., Martin, C., O'Leary, J.J., Ruan, Q., Johnson, D.S., Chen, Y., and O'Neill, L.A.J. (2009). Negative regulation of TLR4 via targeting of the proinflammatory tumor suppressor PDCD4 by the microRNA miR-21. *Nat Immunol* 11, 141-147.

Shortman, K., and Naik, S.H. (2006). Steady-state and inflammatory dendritic-cell development. *Nat Rev Immunol* 7, 19-30.

Soukup, G.A., Fritzsch, B., Pierce, M.L., Weston, M.D., Jahan, I., McManus, M.T., and Harfe, B.D. (2009). Residual microRNA expression dictates the extent of inner ear development in conditional Dicer knockout mice. *Dev Biol* 328, 328-341.

Soumelis, V., and Liu, Y.J. (2006). From plasmacytoid to dendritic cell: Morphological and functional switches during plasmacytoid pre-dendritic cell differentiation. *Eur J Immunol* 36, 2286-2292.

Spits, H., Couwenberg, F., Bakker, A.Q., Weijer, K., and Uittenbogaart, C.H. (2000). Id2 and Id3 inhibit development of CD34(+) stem cells into predendritic cell (pre-DC)2 but not into pre-DC1. Evidence for a lymphoid origin of pre-DC2. *J Exp Med* 192, 1775-1784.

Stefani, G., and Slack, F.J. (2008). Small non-coding RNAs in animal development. *Nat Rev Mol Cell Biol* 9, 219-230.

Steinman, R.M., and Banchereau, J. (2007). Taking dendritic cells into medicine. *Nature* 449, 419-426.

Steinman, R.M., and Cohn, Z.A. (1973). Identification of a novel cell type in peripheral lymphoid organs of mice. I. Morphology, quantitation, tissue distribution. *J Exp Med* 137, 1142-1162.

Steinman, R.M., and Idoyaga, J. (2010). Features of the dendritic cell lineage. *Immunol Rev* 234, 5-17.

Tacken, P.J., de Vries, I.J.M., Torensma, R., and Figdor, C.G. (2007). Dendritic-cell immunotherapy: from ex vivo loading to in vivo targeting. *Nat Rev Immunol* 7, 790-802.

Taganov, K.D. (2006). NF- B-dependent induction of microRNA miR-146, an inhibitor targeted to signaling proteins of innate immune responses. *Proc Natl Acad Sci U S A* 103, 12481-12486.

Thai, T.H., Calado, D.P., Casola, S., Ansel, K.M., Xiao, C., Xue, Y., Murphy, A., Frendewey, D., Valenzuela, D., Kutok, J.L., *et al.* (2007). Regulation of the Germinal Center Response by MicroRNA-155. *Science* 316, 604-608.

Valladeau, J., Clair-Moninot, V., Dezutter-Dambuyant, C., Pin, J.J., Kissenpfennig, A., Mattei, M.G., Ait-Yahia, S., Bates, E.E., Malissen, B., Koch, F., *et al.* (2002). Identification of mouse langerin/CD207 in Langerhans cells and some dendritic cells of lymphoid tissues. *J Immunol* 168, 782-792.

van der Vlist, M., and Geijtenbeek, T.B. (2010). Langerin functions as an antiviral receptor on Langerhans cells. *Immunol Cell Biol* 88, 410-415.

van Niel, G., Wubbolts, R., and Stoorvogel, W. (2008). Endosomal sorting of MHC class II determines antigen presentation by dendritic cells. *Curr Opin Cell Biol* 20, 437-444.

Vigorito, E., Perks, K.L., Abreu-Goodger, C., Bunting, S., Xiang, Z., Kohlhaas, S., Das, P.P., Miska, E.A., Rodriguez, A., and Bradley, A. (2007). microRNA-155 Regulates the Generation of Immunoglobulin Class-Switched Plasma Cells. *Immunity* 27, 847-859.

Villadangos, J., and Heath, W. (2005). Life cycle, migration and antigen presenting functions of spleen and lymph node dendritic cells: Limitations of the Langerhans cells paradigm. *Semin Immunol* 17, 262-272.

Villadangos, J.A., and Schnorrer, P. (2007). Intrinsic and cooperative antigen-presenting functions of dendritic-cell subsets in vivo. *Nat Rev Immunol* 7, 543-555.

Villadangos, J.A., and Young, L. (2008). Antigen-presentation properties of plasmacytoid dendritic cells. *Immunity* 29, 352-361.

Vishwanath, M., Nishibu, A., Saeland, S., Ward, B.R., Mizumoto, N., Ploegh, H.L., Boes, M., and Takashima, A. (2006). Development of Intravital Intermittent Confocal Imaging System for Studying Langerhans Cell Turnover. *J Investig Dermatol* 126, 2452-2457.

von Boehmer, H., and Hafen, K. (1993). The life span of naive alpha/beta T cells in secondary lymphoid organs. *J Exp Med* 177, 891-896.

Vyas, J.M., Van der Veen, A.G., and Ploegh, H.L. (2008). The known unknowns of antigen processing and presentation. *Nat Rev Immunol* 8, 607-618.

Wang, P., Hou, J., Lin, L., Wang, C., Liu, X., Li, D., Ma, F., Wang, Z., and Cao, X. (2010). Inducible microRNA-155 Feedback Promotes Type I IFN Signaling in Antiviral Innate Immunity by Targeting Suppressor of Cytokine Signaling 1. *J Immunol* 185, 6226-6233.

Watowich, S.S., and Liu, Y.J. (2010). Mechanisms regulating dendritic cell specification and development. *Immunol Rev* 238, 76-92.

Wiesen, J.L., and Tomasi, T.B. (2009). Dicer is regulated by cellular stresses and interferons. *Mol Immunol* 46, 1222-1228.

Wilson, N.S., and Villadangos, J.A. (2004). Lymphoid organ dendritic cells: beyond the Langerhans cells paradigm. *Immunol Cell Biol* 82, 91-98.

Wu, C.I., Shen, Y., and Tang, T. (2009). Evolution under canalization and the dual roles of microRNAs--A hypothesis. *Genome Res* 19, 734-743.

Wu, L., and Liu, Y.-J. (2007). Development of Dendritic-Cell Lineages. *Immunity* 26, 741-750.

Xiao, C., and Rajewsky, K. (2009). MicroRNA Control in the Immune System: Basic Principles. *Cell* 136, 26-36.

Xiao, C., Srinivasan, L., Calado, D.P., Patterson, H.C., Zhang, B., Wang, J., Henderson, J.M., Kutok, J.L., and Rajewsky, K. (2008). Lymphoproliferative disease and autoimmunity in mice with increased miR-17-92 expression in lymphocytes. *Nat Immunol* 9, 405-414.

Xu, J., Li, C.X., Li, Y.S., Lv, J.Y., Ma, Y., Shao, T.T., Xu, L.D., Wang, Y.Y., Du, L., Zhang, Y.P., *et al.* (2010). MiRNA-miRNA synergistic network: construction via co-regulating functional modules and disease miRNA topological features. *Nucleic Acids Res* 39, 825-836.

Xu, Y., Zhan, Y., Lew, A.M., Naik, S.H., and Kershaw, M.H. (2007). Differential development of murine dendritic cells by GM-CSF versus Flt3 ligand has implications for inflammation and trafficking. *J Immunol* 179, 7577-7584.

Young, L.J., Wilson, N.S., Schnorrer, P., Proietto, A., ten Broeke, T., Matsuki, Y., Mount, A.M., Belz, G.T., O'Keeffe, M., Ohmura-Hoshino, M., *et al.* (2008). Differential MHC class II synthesis and ubiquitination confers distinct antigen-presenting properties on conventional and plasmacytoid dendritic cells. *Nat Immunol* 9, 1244-1252.

Zenke, M., and Hieronymus, T. (2006). Towards an understanding of the transcription factor network of dendritic cell development. *Trends Immunol* 27, 140-145.

Zhang, D.E., Hohaus, S., Voso, M.T., Chen, H.M., Smith, L.T., Hetherington, C.J., and Tenen, D.G. (1996). Function of PU.1 (Spi-1), C/EBP, and AML1 in early myelopoiesis: regulation of multiple myeloid CSF receptor promoters. *Curr Top Microbiol Immunol* 211, 137-147.

Zhou, J., Cheng, P., Youn, J.I., Cotter, M.J., and Gabrilovich, D.I. (2009a). Notch and wingless signaling cooperate in regulation of dendritic cell differentiation. *Immunity* 30, 845-859.

Zhou, L., Seo, K.H., He, H.Z., Pacholczyk, R., Meng, D.M., Li, C.G., Xu, J., She, J.X., Dong, Z., and Mi, Q.S. (2009b). Tie2cre-induced inactivation of the miRNA-processing enzyme Dicer disrupts invariant NKT cell development. *Proc Natl Acad Sci U S A* 106, 10266-10271.

Zhou, X., Jeker, L.T., Fife, B.T., Zhu, S., Anderson, M.S., McManus, M.T., and Bluestone, J.A. (2008). Selective miRNA disruption in T reg cells leads to uncontrolled autoimmunity. *J Exp Med* 205, 1983-1991.

9 List of abbreviations

AP-1	Activator protein-1
APC	Antigen presenting cell
Batf3	Basic leucine zipper transcription factor, ATF-like 3
BM	Bone marrow
cDC	Conventional DC
CDP	Common DC progenitor
CHS	Contact hypersensitivity
CLP	Common lymphoid progenitor
CLR	C-type lectin receptors
CMP	Common myeloid progenitor
CTL	Cytotoxic T lymphocyte
DC	Dendritic cell
dDC	Dermal DC
Dll	Delta-like
Fig.	Figure
Flt3(L)	FMS-like-tyrosine-kinase 3 (ligand)
GM-CSF	Granulocyte macrophage colony-stimulating factor
Id2	Inhibitor of DNA binding 2
IL	Interleukin
iNOS	Inducible nitric oxide synthase
IRAK1	IL-1R-associated kinase 1
IRF7	Interferon regulatory factor 7
JAK	Janus kinase
ko	Knockout
LC	Langerhans cell
LN	Lymph node
LPS	Lipopolysaccharide
M-CSF	Macrophage colony-stimulating factor
MDP	Macrophage-DC progenitor

MHC	Major histocompatibility complex
miR, miRNA	microRNA
mRNA	Messenger RNA
NF-κB	Nuclear factor-κB
NK cell	Natural killer cell
NOI	Nitric oxygen intermediates
nt	Nucleotide
PAMP	Pathogen-associated molecular pattern
pDC	Plasmacytoid DC
PRR	Pattern-recognition receptor
RISC	RNA-induced silencing complex
RNA	Ribonucleic acid
RNAi	RNA interference
ROI	Radical oxygen intermediates
SHIP1	Src homology 2 domain-containing inositol-5'-phosphatase 1
siRNA	Small interfering RNA
SOCS1	Suppressor of cytokine signaling 1
STAT	Signal transducer and activator of transcription
TCR	T cell receptor
TGF-β1	Transforming growth factor-β1
T_H	T helper cell
TiP DC	TNF/iNOS-producing DC
TLR	Toll-like receptor
TNF	Tumor necrosis factor
TRAF6	TNFR-associated factor 6
T_{reg}	Regulatory T cell
UTR	Untranslated region

i want morebooks!

Buy your books fast and straightforward online - at one of world's fastest growing online book stores! Environmentally sound due to Print-on-Demand technologies.

Buy your books online at

www.get-morebooks.com

Kaufen Sie Ihre Bücher schnell und unkompliziert online – auf einer der am schnellsten wachsenden Buchhandelsplattformen weltweit! Dank Print-On-Demand umwelt- und ressourcenschonend produziert.

Bücher schneller online kaufen

www.morebooks.de

VDM Verlagsservicegesellschaft mbH
Heinrich-Böcking-Str. 6-8   Telefon: +49 681 3720 174   info@vdm-vsg.de
D - 66121 Saarbrücken   Telefax: +49 681 3720 1749   www.vdm-vsg.de

Printed by Books on Demand GmbH, Norderstedt / Germany